Geropsychology and Long Term Care

T0180684

Erlene Rosowsky • Joseph M. Casciani
Merla Arnold
Editors

Geropsychology and Long Term Care

A Practitioner's Guide

 Springer

Editors

Erlene Rosowsky
Harvard Medical School
MSPP, MA
USA

Joseph M. Casciani
Concept Healthcare
San Diego, CA
USA

Merla Arnold
Independent
Geropsychology
Practice
NY, USA

ISBN: 978-1-4419-2483-4   e-ISBN: 978-0-387-72648-9
DOI: 10.1007/978-0-387-72648-9

Printed on acid-free paper

springer.com

Foreword

It is with great pride that the Psychologists in Long Term Care (PLTC) have sponsored *The Professional Educational Long-Term Care Training Manual*, and now its second iteration, *Geropsychology and Long Term Care: A Practitioner's Guide*. Education of psychologists working in long-term care settings is consistent with PLTC's mission to assure the provision of high-quality psychological services for a neglected sector of the population, i.e., residents in nursing homes and assisted-living communities. To this end, direct training of generalist psychologists in the nuances of psychological care delivery in long-term care settings has been a major priority.

It is a tribute to the accelerating nature of research in long-term care settings that a revision is now necessary. After all, the *Professional Educational Training Manual's* initial publication date was only in 2001. However, in the intervening years, much progress has been made in addressing assessment and intervention strategies tailored to the needs of this frail but quite diverse population. It is so gratifying to be able to say that there is now a corpus of scientific knowledge to guide long-term care service delivery in long-term care settings.

Dr. Erlene Rosowsky and Dr. Joe Casciani are to be commended for their hard work in assembling the initial scholar–practitioner team of authors who reveal both a thorough grasp of the long-term care literature and facility in presenting complex material in a readily understandable form. Dr. Robert Intrieri had the task of publishing the first edition. Dr. Merla Arnold has undertaken the task of overseeing its update and of final editing of the manuscripts to assure a high-quality volume. All are accomplished professional educators who have made real contributions to psychology education by enlightening our colleagues about both the art and science of working with this challenging group.

I am quite honored that the initial version of this training manual project was brought to fruition as I completed my tenure as PLTC president. PLTC's spirit and energy continue to offer the most up to date psychological knowledge in long-term care settings to those who work in this exciting field to assure optimal practice with our oldest and most vulnerable citizens.

Tampa, FL   Victor Molinari

Preface

History of the Guide

In the mid-1990s, a number of clinical geropsychologists working with nursing home (NH) residents became aware that the federal agency overseeing health insurance for older adults was taking a close look at, and questioning, psychological services delivered in long-term care settings. This was purported to be in response to their having uncovered some questionable practices among a few clinicians. This was also a time when the demographic shift toward more people living longer and requiring more services was showing impressive growth.

What followed was that 14 seasoned and concerned geropsychologists, all members of Psychologists in Long Term Care (PLTC) and active in their state and national professional organizations, convened to develop a template for standards to describe ethical and clinically sound psychological services in the context of long-term care. This was subsequently published in *The Gerontologist* (1998), 38:1; 122–127.

A strong sentiment emerged from the original work group to develop a program for psychologists, so that their work with older adults was reflective of and consistent with these Standards for Psychological Services in Long Term Care Settings. There was a genuine concern that some good and well-intentioned psychologists, earning their livelihoods working in long-term care settings, might be working in ways that are not consistent with these Standards. In addition, inadequately trained and educated psychologists, working with older adults in long-term care settings, may be insufficiently aware of the special ethical and clinical issues involved in professional work in these settings.

There was an intention to find a way to encourage and help our professional colleagues, not in any way to exclude or discourage them, to deliver high-quality services to many of our nation's most vulnerable, older adults living in long-term care settings. To that end, an educational and training program was developed, as a companion to the Standards. This was then published by a small academic press in 2003 and distributed primarily to the PLTC membership. Subsequently, it was decided to update the material and publish it as this book under the title *Geropsychology and Long Term Care: A Practitioner's Guide*. This expansion was favored over the initial manualized training program at this time, in order to offer this important information, through Springer, to a broader audience. As with the previous edition, for this work we recruited eminent psychologists in each content area to author the specific chapters.

Goals of the Guide

We have endeavored to remain true to the original design in that the chapters of this guide are parallel to and reflective of the units set forth in the Standards.

There is an ever-increasing population of older adults residing in skilled nursing facilities (i.e., NHs) and assisted-living communities. This growth continues to mark a compelling need for skilled professionals with proficiency in the fields of geropsychology and in aging and mental health, to meet the needs of older adults not only right now, but also in the future.

There are as many as 1.5 million people living in NHs across the United States, according to the Surgeon General's Report (1999). Although up to two-thirds of these NH residents were noted to have a diagnosable mental disorder, these conditions remain largely underdiagnosed and undertreated. In our effort to assist psychologists and others fill this serious gap in care, we offer this text. Knowing the landscape of long-term care settings and being aware of the special needs and issues facing older adults who live there will help clinicians address the residents' often unmet mental and behavioral health needs, with its serious concomitant risk of untoward cascading effects. Thus, we offer the material in this guide as a piece of the foundation of high-quality practice, when working with older adults living in long-term care settings.

An additional goal of this work is to help generate greater awareness of the knowledge base and skills that are important in this APA proficiency area. Understanding federal policy, the types of patients who may be referred, and the various credentials of the varied providers involved in this work, and the importance of comprehensive assessment, treatment planning, delivery of services, and medical records documentation remain at the top of the list of professional skills that are essential to this work. Furthermore, providers must be well apprised of the ethical issues and guiding professional principles inherent in this work.

Content of the Guide

There are eight chapters to this text. Chapter 1, authored by Dr. Norris, addresses the incidence of mental illness in long-term care, the policies affecting the delivery of mental health services, and reimbursement issues. Chapter 2 is authored by Dr. Smith. He discusses funding issues and how the referral process often unfolds in NHs, followed by a discussion of the kinds of referrals that can be expected, along with some of the challenges that come with these varied referral types. In Chap. 3, Drs. Edelstein, Northrop, and MacDonald cover, in a very thoughtful manner, the multidimensional and multi- and interdisciplinary nature of assessment with older adults. Included is a review of relevant psychological instruments, as well as the many issues important to consider when conducting "responsible assessments" with frail older adults living in long-term care settings.

Chapter 4, written by Dr. Frazer, focuses on treatment plans, their intrinsic qualities, and how they are developed. In Chap. 5, Dr. Duffy explores several facets of treatment, including specific treatment approaches having demonstrated effectiveness with older adults. There is a discussion of the importance of integrating mental and behavioral health services with multiple disciplines and people who are also responsible for providing services. Dr. Duffy also points out that, when warranted and appropriate, discontinuing mental health services is an important element in the treatment process, worthy of special consideration. A new chapter in this edition, Chap. 6 by Drs. Hyer and Shah, introduces an important element in the training of and services delivered by geropsychologists, namely the integration of psychiatric medication understanding with consultation and psychological services.

In Chap. 7, the authors Drs. Arnold, Colletti, and Stilwell, draw contrasts among the multiple disciplines providing services, including mental health services, in long-term care settings. To work effectively on an interdisciplinary team and to provide the highest quality of services, it is important to know about the roles and responsibilities of others providing health care services in long-term care settings. The discussion includes how the mental health disciplines differ from each other with respect to levels of training required of those proficient in geropsychology. There is an important focus on the topic of documentation from the perspectives of the Medicare carriers, privacy concerns, and outcomes measurement.

Lastly, in Chap. 8, Dr. Karel takes up the topic of ethical issues, including consent to treatment, the right to privacy and confidentiality. Dr. Karel offers an important framework for ethical decision-making in long-term care.

Concluding Thoughts

This new publication, we believe, represents some of the best thinking in the field. We anticipate that this text will be a useful guide and educational resource to psychologists and other mental health practitioners working in, or endeavoring to work in, long-term care settings. It is the final product of many years of planning, discussion, and writing, from its inception in the mid-1990s to the publication of this second version in 2008. A publication like this can only come from the contributing authors' dedication and many hours of effort, and we want to thank all of them for their dedication and scholarly work. Of particular note, the authors have generously agreed to allow PLTC to receive all proceeds earned from this project. For that, we are very grateful.

To the extent that this text raises the skill level and knowledge base of our colleagues, and better prepares them for this exciting and demanding specialty practice area of geropsychology, the goals of the editors and PLTC will have been well met.

Boston, MA   Erlene Rosowsky
San Diego, CA   Joseph M. Casciani
Huntington Station, NY   Merla Arnold

Contents

Foreword v

Preface vii

1 **Policies and Reimbursement: Meeting the Need
for Mental Health Care in Long-Term Care** 1
Margaret Norris

2 **Referrals** 13
Michael Smith

3 **Assessment** 23
Barry A. Edelstein, Lynn E. Northrop, and Lisa M. MacDonald

4 **Treatment Plans** 49
Deborah W. Frazer

5 **Treatment Process** 55
Michael Duffy

6 **Integration of Psychology, Psychiatry, and Medication
in Long-Term Care** 65
Lee Hyer and Shalija Shah

7 **Professional Practice: Disciplines, Documentation,
and Outcomes Measurement** 87
Merla Arnold, John C. Colletti, and Nicholas C. Stilwell

8 **Ethical Issues in Long-Term Care** 111
Michele J. Karel

Index 125

Chapter 1
Policies and Reimbursement: Meeting the Need for Mental Health Care in Long-Term Care

Margaret Norris

Almost 40% of older adults will spend some portion of their lives in a long-term care (LTC) facility (Seperson, 2002). For reasons outlined in this chapter, mental health services are in high demand in LTC settings. This chapter briefly summarizes the demographics of the LTC population and their mental health needs. The chapter also reviews the public policies and reimbursement parameters that make psychology services in LTC viable and rewarding.

Mental Illness in Long-Term Care Residents

The proportion of nursing home (NH) residents with mental health problems is astounding. Estimates of mental illness, including dementia diagnoses, are roughly 65-80% (Burns et al., 1993). Excluding those with dementia, the prevalence rate is ~20%, which remains far higher than seen in community-dwelling populations. However, an insufficient number of residents are receiving mental health care. One study reported that less than 20% of those diagnosed with a mental illness in a NH received treatment for that disorder in the previous year (Smyer, Shea, & Streit, 1994).

There are obvious reasons why mental illness is so prevalent in NHs. First, there is a strong relationship between physical and mental health problems. Depression, anxiety, and behavioral disturbances can be exacerbated by pain, diabetes, stroke, hip fracture, and other medical illness that commonly precipitate long-term care (LTC) placement. Similarly, mental health problems can exacerbate physical suffering (e.g., pain, obesity) and interfere with motivation and effort in rehabilitation therapies.

Dementia, particularly in the more moderate to severe stages, can cause behavior problems that are often difficult to manage at home, resulting in the need for 24 hour care and can precipitate a move into an LTC setting. Placement is frequently accompanied by additional losses and associated transitions such as decline in functional independence, loss of home and personal possessions, loss of family and neighbors, chronic and/or terminal health conditions, and adjustments to staff, roommates, and often a myriad of institutional rules.

E. Rosowsky et al. (eds.), *Geropsychology and Long Term Care*,
DOI: 10.1007/978-0-387-72648-9_1, © Springer Science+Business Media, LLC 2009

LTC may be delivered in a variety of settings, including NHs, assisted-living facilities or communities, adult day care facilities, personal care homes, rehabilitation units, and even in community settings, at the person's home. The common feature is that a range of health care services, including medical, personal, social, and psychological services is delivered to meet the needs of people with disabilities, chronic illnesses, or limitations in their ability to function independently. A recent trend is a decrease in the proportion of older adults placed in NHs, with a larger proportion going to assisted-living facilities (Cohen et al., 2003). The fastest growing segment of residential facilities for older adults is assisted living. The estimated number of assisted-living facilities in the United States is approximately 33,000 (Cohen et al.). As with NHs, there is a considerable need for mental health services in assisted-living settings because cognitive impairment and behavioral disturbances are often precipitants to placement in both NHs and assisted living, as noted earlier. The primary difference in these two populations is medical status, with NH residents being more ill and frail. The similarities, however, are substantial. Residents of both NHs and assisted-living have higher rates of mental illness and behavior problems than that occur in community-dwelling older adults. As detailed below, there tends to be important differences in the funding of mental health care for NH and assisted-living residents.

The rapidly changing ethnic make-up of the older adult population in the United States is another important demographic shift. By 2050 for example, the percentage of the Caucasian segment of the older adult population is expected to decline from 84% to 64% (Seperson, 2002), with a corresponding rise in the numbers of more diverse segments of the population older adults. This growth in minority populations will be reflected in the LTC population. Hence, mental health services in LTC facilities must accommodate ethnically and culturally diverse populations.

Based on 1990 U.S. Census data on persons aged 60 and above, a study showed that the rates of NH use was 3.3% for Whites, 3.1% for Blacks, 2.3% for Native Americans, 1.6% for Hispanics, and 1.2% for Asians (Himes, Hogan, & Eggebeen, 1996). Lower rates for Asians, Hispanics, and Native Americans may reflect cultural differences in family caregiving patterns, as well as language barriers, or limited access to formal services. Even the need for mental health services may vary according to ethnicity. For example, Weintraub et al. (2000) found the rate of dementia on admission to NHs higher among Black residents than among White residents. In addition, minority demographics differ substantially across regions of the United States. Psychologists in different localities may have a large proportion of first- or second-generation immigrants from Southeast Asia, China, Korea, Cuba, Puerto Rico, Eastern Europe, Russia, and so forth. It is imperative that psychologists working with these populations become familiar with their cultural values and attitudes, particularly about germane matters such as mental illness, health care, and familial caregiving traditions.

Psychologists consulting in LTC must not only be competent to serve diverse populations, they must also stretch their familiar role as geropsychologists. LTC placement is rapidly growing among younger persons because of increased survival rates among those with traumatic injuries and illnesses. Younger residents often present with significant emotional difficulties such as depression, resulting from the dramatic changes in their lives brought on by disabilities, as well as behavioral disturbances, as they try to fit into an institutional life that historically has been aimed toward frail older adults.

LTC psychologists must adopt principles of respect and recognition of the uniqueness of different cultural identities and be prepared to work in an informed, culturally-competent manner. Looking at the bigger picture, psychologists may also play a valuable role in LTC by educating and training LTC staff, such as nurses and nursing aides, helping them to be more aware of and comfortable with these ethnic and cultural variations.

Policies Impacting the Delivery of Mental Health Services in LTC

OBRA Increased Access to Mental Health Services. Prior to 1989, the financial survival of psychologists working in LTC settings was tenuous at best. In recent years, public policies have expanded access to mental health services for older adults. Mental health services in LTC are now mandated by federal law under defined circumstances. Further, Medicare reimbursement rates and those of traditional fee-for-service insurance plans are currently at or above the fee schedules offered by most HMOs and managed care plans; however, there has been a downward trend in this regard. Serious limits to the access and reimbursement of mental health services remain, particularly for NH residents. Both these expansive and the restrictive policies are briefly reviewed below.

Regulation of NHs has been established primarily through the Omnibus Budget Reconciliation Acts (OBRAs) of 1986, 1987, 1989, and 1990. These OBRA laws also addressed major reimbursement issues relevant to all Medicare beneficiaries, regardless of residence. The most comprehensive reforms came from OBRA 1987, which greatly impacted the access of mental health care for older adults. The previous $250 annual cap on outpatient mental health care allowed by Medicare was increased to $450 and later to $1,100. OBRA 1989 had the next major impact on providing mental health services to older adults by repealing Medicare's cap on mental health care. OBRA 1987 also removed the restriction on settings where psychologists were allowed to provide services; thus, opening the opportunity for residents to receive psychological services in LTC settings. In addition, OBRA 1987 mandated that (1) all NH residents be screened to determine if they have a mental illness or mental retardation and whether they need treatment; (2) in-services training on mental health services be increased; (3) "chemical agents" and physical restraints be restricted; and (4) the cognitive, behavioral, and psychiatric functioning of all residents be documented with standardized assessment methods.

The unsurprising result of OBRAs 1987 and 1990 was large increases in coverage for mental health services to Medicare beneficiaries. From 1991 to 1993, reimbursements increased 53% in all settings and increased 244% in NH settings. A fourfold increase in NH mental health services eventually became the basis for the

Department of Health and Human Services Office of the Inspector General's (OIG's) first major investigation of these services in NHs. There was a serious oversight in the basis for this investigation. While four times as many mental health services were being provided in NHs, the rate of mental health disorders in NHs is six times the rate found in the community. This difference initially suggests NH residents were underserved; however, as reviewed below, a pattern of inappropriate services also was observed.

OIG Investigations of Fraud. Psychologists providing services in LTC settings are strongly advised to pay attention to OIG reports. The OIG is mandated to protect the integrity of the programs provided by the Department of Health and Human Services and the welfare of the beneficiaries served by them. The OIG mission is carried out by conducting audits, investigations, inspections, applying sanctions, and generating fraud alerts. The OIG may recommend legislative, regulatory, and operational approaches to address problems. Four reports concerning Medicare mental health services in NHs have been published in recent years. These include Mental Health Services in Nursing Facilities (May 1996), Medicare Payments for Psychiatric Services in Nursing Homes: A Follow Up (January 2001), Psychotropic Drug Use in Nursing Homes (November 2001), and Medicare Carriers' Policies for Mental Health Services (May 2002). OIG reports can be accessed at http://www.hhs.gov/oig/oei.

The first report published in 1996 uncovered serious problems. The report concluded that 32% of mental health services in NHs were medically unnecessary and another 16% of services were deemed questionable because of inadequate documentation. Services most often found to be inappropriate were psychological testing, group therapy, therapy provided to residents with advanced dementia, medication management billed as psychotherapy, and "incident to" services provided by either unidentified or unqualified persons and billed under a licensed practitioner. The report also acknowledged another side to the coin, stating "We found that some beneficiaries are not getting the care they need" (p. 10). Seventy-eight percent of the NH respondents (administrators, directors of nursing, and social workers) reported that barriers to receiving mental health services still existed in NHs. Further, 70% felt that the inclusion of psychologists as Medicare providers was beneficial to NH residents, primarily because psychologists were more available than other specialists to come to NHs. Finally, the OIG determined that Medicare carriers were also culpable because of inadequate policies for mental health services.

The OIG follow-up study 5years later, in 2001, concluded that inappropriate mental health services declined from the previous rate of 32% to 27%, representing a small improvement and still a serious concern. Parenthetically, this rate stands in contrast to 15% of psychotropic medication use found to be inappropriate by the OIG in the November 2001 report. The transgressions identified included providing services to residents with significant cognitive limitations and providing therapy services with unjustifiable durations and frequencies. In addition, the report concluded that most Medicare carriers had since added mental health policies, but utilization guidelines remained imprecise.

The OIG Medicare Carriers' Policies for Mental Health Services (May 2002) investigation reviewed the Medicare carriers' criteria and documentation requirements for the most common mental health procedures. The report concluded that there was a great deal of variability in the comprehensiveness and specificity of the carriers' local policies on mental health services. For example, utilization guidelines for individual therapy, group therapy, psychological testing, and medication management were inconsistent across carriers. These inconsistencies create a great deal of disparity in the mental health services allowed for NH residents, depending on which carrier exists in their state.

In an invited comment to the report, the American Association for Geriatric Psychiatry (AAGP) cogently pointed out that the quality and comprehensiveness of mental health services should not vary depending on which carrier is reimbursing for the service. Their comment also strongly argued that allowing carriers to develop utilization guidelines allows for potential discriminatory practices based on payment restrictions rather than scientific knowledge and clinical practices. In conclusion, psychologists must become familiar with the Local Coverage Determinations (LCDs) in order to know the regulations that determine what services are reimbursable by their Medicare Administrative Contractors (MACs) (formerly known as "carriers"). MACs can be found at http://www.cms.hhs.gov/MedicareContractingReform/.

As reviewed earlier, reimbursement for mental health services has improved tremendously since OBRAs 1987 and 1990. Services have greatly increased, as expected and as indicated from the documented need for services. However, psychologists must heed the findings of the OIG in order to identify high-risk areas of their practices in LTC settings. Despite the improved reimbursement regulations, other policies greatly impact the access to services, albeit in a negative direction. These include the mental health outpatient limitation applied to mental health services by Medicare, as described below, and the limits applied at the state level to Medicaid coverage of mental health services.

Reimbursement Systems of Mental Health Care in LTC

Outpatient Mental Health Treatment Limitation. Medicare reimburses 80% of the approved amount for essentially all outpatient services, leaving a remaining 20% copayment. Psychotherapy is an exception. Known as the "outpatient mental health treatment limitation," Medicare reduces payment for all outpatient therapy services by 62.5%, effectively reducing the payment amount from 80% to 50% of the allowed amount. This is illustrated in the following example:

Medicare approved amount=$100
Outpatient treatment limitation ($100×0.625)=$62.50
80% of limitation amount ($62.50×0.80)=$50

Hence, Medicare pays for 80% of 62.5% of the approved amount, which results in a 50% reimbursement rate. This limitation shifts the responsibility of 50% of the approved amount from Medicare to the patient and any secondary insurance that the patient may have. It is important to note that this limitation rule applies to outpatient psychotherapy services, not to diagnostic services (including the initial diagnostic interview, procedure code 90801). It also does not apply to inpatient therapy services, or to psychological testing. An important inconsistency is that it does apply to therapy services provided in NHs, despite the fact that the therapy procedure codes used for services in NHs are inpatient codes (e.g., 90816, 90818).

A major victory was achieved in 2008 when the Medicare Improvements for Patients and Providers Act was passed. This law will gradually phase out the therapy limitation. Beginning in 2010, the patient copay will be reduced every year until 2014 when the copay will be the same 20% that exists for all outpatient services. This law will eliminate one of the most serious barriers to mental health services for older adults. Until it is fully in effect, providers must be aware of the implications of the therapy limitation.

The outpatient mental health therapy limitation places a great deal of emphasis on patients' secondary insurance policies. NH and assisted-living residents tend to be affected in different ways. Approximately 70% of NH residents are "dually eligible," i.e. they have both Medicare and Medicaid insurance (AARP Public Policy Institute, 2001). As a federal-state insurance program, Medicaid policies are determined by both the federal and state governments. By federal law, mental health services are defined as ancillary services, giving the individual states a great deal of latitude in determining what services will or will not be reimbursed. Approximately 75% of the states do not allow for "crossover" payment, i.e., reimbursement of copayment of mental health services. Hence, the vast majority of NH residents do not have coverage for 50% of the cost of mental health services, rendering it largely unavailable to most NH residents in states that do not have crossover, or requiring the mental health practitioner to write off the uncollectible amount.

In contrast, the majority of assisted-living residents are not eligible for Medicaid. Less than 10% of assisted-living residents receive Medicaid benefits (Cohen et al., 2003). Approximately 75% of these residents pay for their stay with their own funds and they typically have private secondary health insurance policies. These policies typically pay all or a portion of the patient's copayments for health care expenses, including those for psychotherapy. Some policies are "Medigap" secondary insurance policies. These are regulated by the federal government and the 50% copayment must be covered by the Medigap insurance. If the secondary insurance is not a Medigap policy, all or a portion of the 50% copayment may be covered, and the patient is responsible for any remaining balance. It is critical to note that routinely waiving copayments and deductibles is illegal. Providers must attempt to collect these payments. Under circumstances of financial hardship, and with appropriate documentation, the copayment may be waived.

Medicare Administrative Contractors. The Centers for Medicare and Medicaid Services (CMSs), within the Department of Health and Human Services, administers the Medicare program. CMS contracts with health care insurance agencies to administer the Medicare program in local areas. The insurance agencies that handle Medicare Part A and B claims are known as "Medicare Administrative Contractors (MACs)."

The companies that process claims for Part A of Medicare are referred to as Fiscal Intermediaries. CMS establishes national Medicare policies or, National Coverage Determinations, which all MACs must follow. The MACs are also given leeway to establish additional policies known as LCDs. LCDs across MACs vary on important dimensions such as utilization guidelines, covered diagnoses, and documentation requirements. Because of differences in LCDs across MACs, it is critical for individual providers to be familiar with the LCDs of their respective MAC. LCDs are available on the individual MAC Web sites; updates are published in MAC newsletters; and proposed changes to LCDs must be made available to providers in the community for comments. Contact information for all MACs can be found at http://www.cms.hhs.gov/MedicareContractingReform/. Providers are held accountable for adhering to LCDs. As postpayment audits are not uncommon and should be expected if claims suggest that a provider is not adhering to the MAC LCDs; psychologists should be thoroughly familiar with their MAC's LCD for psychiatry or psychology services.

Consultation Requirement. When psychologists become Medicare providers, they sign an agreement whereby they agree to consult with the patients' primary care or attending physician. Patients are permitted, of course, to give or withhold consent for this consultation. If a patient consents, the psychologist is required to consult "within a reasonable amount of time." The extent of the consultation is determined by the providers. If a patient has been referred to the psychologist by the patient's physician, then the consultation is presumed to have occurred. This rule is relevant to LTC settings, especially NHs, where physician orders for psychology services is typically required. Thus, the physician is aware that his or her patient will be receiving psychological services. Nevertheless, it is prudent for psychologists in LTC settings to communicate with physicians following their initial evaluation of patients and inform physicians of their treatment plans.

If the patient declines to give consent, the psychologist should document that the patient does not want the consultation to occur. Again, this rule has particular meaning for LTC residents. This documentation is necessary but, if it is placed in the patient's medical chart at the facility, the physician will have access to the information that the patient has declined to give to them.

Billing and Coding. Claims for Medicare (and virtually all insurance companies) must be submitted on the CMS Form 1500. These may be submitted by paper or electronically. Electronic billing is processed and paid much more quickly than paper forms. In addition, errors on electronic forms are quickly returned for corrections, again resulting in faster payment. Further, payments can be made either by check or by electronic transfer of funds into the provider's bank account.

Psychologists are generally advised to bill "actual amounts," that is, the fee they establish for their service. Medicare's payment, however, will be based on their "approved amount" for their service. Although it is important that psychologists do not routinely lower their fees to match that paid by an insurance company, psychologists must also know that as a Medicare provider, they have agreed to not bill patients for the difference between the actual amount and the approved amount. This is referred to as "accepting assignment" and is a regulation that applies to all provider agreements that psychologists have with Medicare.

There are several coding issues on the claim forms that must be performed correctly or the claim will be denied. These include place of service, procedure code, and diagnosis codes. The claim must indicate where the service took place. Places of service may include skilled NH, assisted-living facility, patient's home, outpatient office, and so forth.

Current procedural terminology (CPT) codes define the particular service that was provided to the patient. Therapy procedure codes designate whether the service was in an inpatient or outpatient setting, and the length of the session. For example, 90816 indicates a brief (20-30min) inpatient psychotherapy session; 90818 is used for a standard length (45-50min) inpatient (NH) psychotherapy session; 90806 is a standard length outpatient psychotherapy session. A common reason for denial of a claim is that the procedure code and place of service are contradictory, one indicating an outpatient service and the other indicating an inpatient place of service. Diagnostic and testing codes, such as 90801 and 96101 respectively, are not defined according to time or place.

Diagnostic codes must also be reported on the claim form. These designate the condition for which the patient is receiving service. Diagnostic codes are critical for establishing the medical necessity of the service. Medicare requires that diagnostic codes be based on the International Classification of Diseases, 9th Edition (ICD-9). However, many psychologists are inclined to use DSM diagnostic codes. Although codes from the ICD-9 and the DSM mostly correspond with each other, that is not always the case. ICD-9 codes may have a higher level of specificity, which Medicare requires. In addition, ICD codes are more extensive than DSM codes. This broader range of codes to select from is especially important for testing services and for health and behavior intervention and treatment (H&B) services. An excellent coding reference, listed in the Resources section at the end of this chapter, is Easy Coder: Psychiatry/Psychology (Tanaka, 2003). This book also lists medical conditions that would be common diagnoses for patients receiving Health and Behavior services.

Excluded Services. National and local "screens" are used to identify claims that may be denied on the basis of medical necessity. Local screens are more difficult to identify because they vary across MACs and are subject to change over time. There are MACs that provide information about their screens. For example, a common screen is the frequency of psychotherapy sessions. Some MACs may request documentation to support the medical necessity of psychotherapy extending beyond 20 sessions per year. Other screens have been disallowed by CMS, including automatic denials of psychotherapy claims for patients with a diagnosis of dementia. Although MACs may no longer use a dementia diagnosis as a screen, they can request documentation to substantiate that a patient with dementia is benefiting from services.

Medicare pays for services that are "reasonable and necessary for the diagnosis or treatment of an illness or injury or to improve the functioning of a malformed body member." Medicare does not reimburse for preventive services, screening procedures, or routine reassessments. Hence, screening for mental illness of new admissions in an LTC setting is not reimbursed. Testing current patients on an annual basis without clinical justification or "medical necessity" would likely be considered unnecessary, and therefore, not billable.

Family psychotherapy is covered by Medicare but only when treatment of the patient's condition is the primary purpose for the therapy. For example, Medicare would reimburse when there is a need to observe the patient's interaction with his or her family, the family needs assistance in caring for the patient, or the family's capacity to assist the patient needs to be evaluated. Medicare does not reimburse for services directed at the effect of the patient's condition on the family. Family psychotherapy is billed as code 90847, indicating the patient was present at the session. Some, but not all, MACs will also reimburse for procedure code 90846 in which the family was seen without the patient present.

Conclusions

The demand for mental health services in LTC settings is enormous. Recent laws have greatly improved access of these services, rendering LTC work a viable option for psychologists. As with any specialty area, psychologists must become familiar with the specific nuances that determine the success of that work. LTC psychologists must understand the unique needs of this population through training, education, and consultation, such as that provided by this text. In addition to the myriad of clinical issues, psychologists must gain knowledge of practical matters such as laws and regulations that determine the delivery and reimbursement of their services. Finally, as individuals and member organizations, psychologists play a key role in advocating for LTC residents to help ensure that older adults receive the mental health services they need. For further information on advocacy work, visit the PLTC Web site at http://www.wvu.edu/~pltc and click on the Public Policy section.

Key Points

1. There is a great need for mental health services in LTC because of the high rates of emotional and behavioral disturbances that are associated with health decline and major losses.
2. Most LTC services are provided in NHs and assisted-living facilities. Other LTC settings include adult day care centers, rehabilitation units, personal care homes, and even the older adult's home.
3. Psychologists should be prepared to work with a heterogeneous population in LTC, including diverse ethnic, income, and age groups. Psychologists may also play a key role offering staff training on cultural diversity.
4. Older adults' access to mental health services was greatly increased by OBRA laws that allow psychologists to practice in all LTC settings, remove previous caps on mental health coverage, and mandate more services to meet the mental

health needs of NH patients. These policy changes resulted in substantial increases in Medicare payments for mental health services, especially in NHs.

5. As a result of the rapid increase in mental health costs for NH residents, the OIG conducted a series of investigations, the results of which psychologists in LTC settings should heed. LTC services that remain under scrutiny are group psychotherapy, psychological testing, therapy provided to residents with advanced dementia, and "incident to" services. The OIG reports also recognized that psychological services are beneficial to residents and many residents still do not have adequate access to mental health services.

6. Medicare's outpatient mental health treatment limitation results in a 50% copayment for psychotherapy services, which will be gradually phased out until 2014. This copayment may be covered in full or in part by the patient or his or her secondary insurance.

7. The vast majority of NH residents' secondary insurance is Medicaid. Medicaid payment of the 50% copayment varies across states, with only a small number covering this crossover payment.

8. The vast majority of assisted-living residents have private secondary health insurance policies, most of which pay the 50% copayment in part or in full.

9. Psychologists in LTC settings must become familiar with their local MAC's LCDs because these determine what services are, and are not, reimbursable.

10. As Medicare providers, psychologists must agree to consult with patients' primary care physicians, unless the patient does not consent to the consultation. In LTC settings, where physicians typically write orders for psychological services, psychologists should have a method for communicating their initial evaluations and treatment plans to physicians.

11. In filing claims for reimbursement, psychologists must be familiar with Medicare's approved fees and coding of place of service, procedure codes, and diagnostic codes.

12. Psychologists should become familiar with services that are excluded from reimbursement. National exclusions or "screens" can be identified, such as preventive services or routine reassessments. Local screens also exist and are most often identified in MAC LCDs.

References

AARP Public Policy Institute. (February, 2001). *Nursing homes: Fact sheet.* http://www.aarp.org/research/longtermcare/nursinghomes/aresearch-import-669-FS10R.html.

Burns, B., Wagner, H. R., Taube, J. E., Magaziner, J., Permutt, T., & Landerman, L. R.(1993). Mental health service use by the elderly in nursing homes. *American Journal of Public Health*, 83, 331–337.

Cohen, G. D., Blank, K., Cohen, C. I., Gaitz, C., Liptzin, B., & Maletta, G. (2003). Mental health problems in assisted living facilities: The physician's role in treatment and staff education. *Geriatrics*, 58(2), 44, 54–55.

American Psychological Association. (2000). Medicare handbook: A guide for psychologists. http://www.apapractice.org.

Seperson, S. B.(2002). Demographics about aging. In S. B.Seperson & C. Hegeman (Eds.), *Elder care and service learning: A handbook*. Westport, CT: Auburn House.

Himes, C. L., Hogan, D. P., & Eggebeen, D. J. (1996). Living arrangements of minority elders. *Journal of Geronotlogy, 51B*, S42–S48.

Smyer, M. A., Shea, D., & Streit, A. (1994). The provision and use of mental health services in nursing homes: Results from the national medical expenditure survey. *American Journal of Public Health*, 84(2), 284–287.

Tanaka, P. K. (2003). *Easy coder: Psychiatry/psychology*. Montgomery: Unicor Medical Inc.

Weintraub, D., Raskin, A., Ruskin, P. E., Gruber-Baldini, A. L., Zimmerman, S. I., Hebel, J. R., et al. (2000). Racial differences in the prevalence of dementia among patients admitted to nursing resources homes. *Psychiatric Service*, 51, 1259–1264. Resources

MAC draft LCDs:http://www.cms.hhs.gov/mcd/

Medicare Administrative Contrators list:http: //www.cms.hhs.gov/MedicareContractingReform/

Office of the Inspector General Reports: http://www.hhs.gov/oig/oei

Psychologists in Long-Term Care Web site: http://www.wvu.edu/~pltc

Chapter 2
Referrals

Michael Smith

Introduction

This chapter will discuss both the process of referral in long-term care settings and the kinds of referrals psychologists typically get in these settings. The emphasis will be on nursing homes, but mention will be made of assisted-living facilities, retirement residences and other settings, especially when there appear to be differences between these settings and nursing homes.

The Referral Process

The process of referral is complex for at least two reasons. First, different insurance carriers have different requirements and procedures for referral; second, nursing homes and, to a lesser extent, other long-term care settings are heavily regulated by state and federal law.

Insurance

The primary insurer for most residents of long-term care is Medicare, which includes coverage for psychological services under Part B. Depending on the local carrier, there is some variability in how Medicare Part B is administered across the country. Carriers are private insurance companies that have contracted with Medicare to handle claims for a given area, usually one or more states. Under general Medicare guidelines, each of these carriers puts into place its own policies regarding coverage for services, so-called local coverage determinations (LCDs). Each of these carriers has (or should have) issued an LCD for mental health services and may have issued updates (revisions to LCDs) as well. Currently (2008–2009), this carrier-based system is being revised. The country is being divided into 15 different jurisdictions, groups of 2 or more states or territories (Puerto Rico, Virgin Islands, and other territories are included). New carriers, each

E. Rosowsky et al. (eds.), *Geropsychology and Long Term Care*,
DOI: 10.1007/978-0-387-72648-9_2, © Springer Science+Business Media, LLC 2009

one to now be called a Medicare administrative contractor (MAC), are being selected for each jurisdiction.

To be eligible to file claims and be reimbursed under Medicare, psychologists must apply and be credentialed by Medicare as providers. Before 1989, psychologists could be Medicare providers and receive reimbursement for certain diagnostic services, but only under limited circumstances and requirements. One such requirement was that all patients receiving psychological services had to be referred by a physician. In 1989, a new federal law that expanded the psychological services covered under Medicare Part B (the Omnibus Budget Reconciliation Act of 1989 [OBRA-89]) was passed. This law and the regulations putting it into effect in 1990 expressly prohibited the physician referral requirement. Instead, psychologists providing diagnostic or therapeutic services were required to notify the patient's physician that psychological services were being provided, unless the patient specifically requested that notification not be done.

A peculiar result of these pre- and post-1989 laws and regulations is that both sets of regulations still exist – the post-1989 rules did not replace the pre-1989 ones. They were simply added to them. As a result, psychologists may be accepted as Medicare providers under either set of rules. Psychologists accepted under the pre-1989 rules are typically designated as "psychologists" (and have a "provider code" of 62) under Medicare nomenclature, while psychologists accepted under the post-1989 rules are designated as "clinical psychologists" (and have a "provider code" of 68). A few psychologists, for various technical or financial reasons, still choose to enroll with Medicare as "psychologists" (62), but the vast majority of psychologists enrolled with Medicare are "clinical psychologists" (68) and are thus exempt from any physician referral requirement for Medicare reimbursement.

For long-term care residents whose primary insurer is not traditional Medicare, referral requirements may be different. These residents may be enrolled in a Medicare HMO, or have other primary non-Medicare insurance. Medicare HMOs and these other insurers typically have a variety of referral requirements, including referral from the patient's primary care provider or preauthorization for mental health services. These requirements can be complex, frequently change, and can vary according to the setting the patient is in, whether the psychologist is enrolled as a provider with the particular insurer in question, and other factors.

State Laws and Regulations

State laws and regulations obviously vary from state to state, but nursing homes are usually heavily regulated, much more than other long-term care settings, which may or may not be regulated at all. Nursing homes, following regulations from the state health department or similar regulatory body, generally are required to use a hospital model where each patient has an attending or primary physician who must order services of all other physicians and providers. Typically then, psychologists practicing in nursing homes, along with all other providers, must receive referrals via a written order from the attending physician. In the attending physician's absence, the covering physician may write orders, and the medical director of the nursing home generally has this authority as well.

In other long-term care settings such as assisted-living and retirement residences, there is often little or no state regulation. When there is, these settings are often not considered healthcare facilities to be regulated by a health department, but are instead monitored or overseen by a state social service department or similar agency. In particular, they are rarely considered organized healthcare settings with patients assigned to attending physicians and with a protected, legally regulated medical record. Physician orders or referrals do not apply, and residents of these settings are free to choose providers and self-refer just like anyone else in the community.

These differences in state regulation of nursing homes vs. other long-term care settings have implications for HIPAA requirements as well. For example, patient medical records in a nursing home are like patient records in a hospital. It is the legal responsibility of the nursing home to protect and to release when appropriate. The documentation placed in the medical record by psychologists and other consultant healthcare providers who see patients in the nursing home is part of that record. Copies of this documentation are typically provided to these consultants as a courtesy, to facilitate ongoing treatment and insurance billing, but the responsibility for control and disposition of the record remains with the nursing home. Thus, in this arrangement, the nursing home should distribute a HIPAA privacy notice that applies to their medical records and covers all the providers who practice there, and these consultant providers should refer to the nursing home any requests they receive for release of medical records.

In contrast, other long-term care settings rarely, if ever, have a legally regulated medical record, and thus the consultant providers' records are their own and the HIPAA rules governing privacy notices, maintenance of records, release of information, etc., apply just as they do to private office practice.

Informal Aspects of the Referral Process

The foregoing applies to the formal aspects of the referral process, but of course, there are informal aspects as well. In nursing homes, for example, where a physician's written order is required, the impetus for a referral can come in many ways. In some cases, the physician may initiate the referral, but in many cases it may come at the request of someone else – nursing staff, the social worker, recreation staff, rehabilitation staff, another consultant (e.g., a psychiatrist or physiatrist), the patient's family, or the patient. An individual staff member who sees the need for a referral may approach the physician directly, or the need for a referral may emerge at a staff meeting.

Decision-making networks and patterns of social influence vary in different nursing homes. In some homes, physicians may be heavily involved in patient care, while in others physicians may not be on site more than a few hours a day and may be less involved. Where they are less involved, nurses may play a greater role in influencing patient care; in other cases social work staff may figure prominently in initiating referrals to psychology. In long-term care settings other than nursing homes, referrals may also be initiated in a variety of ways. Depending on how a particular facility is organized and administered, there may be an executive director, a nursing director, a case manager, or other staff members who initiate referrals. Families and patients themselves may play a greater role. A given facility may have set a policy that it follows.

In addition to all of this institutional variation, individuals vary in how readily they recognize when mental health services and psychological services, in particular, are needed. Physicians and other staff may vary in how sensitive they are, for example, to signs of depression in patients (depression is notoriously undertreated in older people). Those low in sensitivity may simply fail to notice the signs. Or, they may notice the signs and recognize the depression, but dismiss it as expected or "normal" and thus somehow not worthy of treatment. Or, they may agree that treatment is needed but decide to treat only with medication and fail to appreciate the value and effectiveness of psychotherapy.

The use of psychotropic drugs to treat problems, to the frequent exclusion of alternative or complementary psychological interventions, deserves expanded discussion. In nursing homes, since the late 1980s, federal regulations have discouraged the use of psychotropic drugs, except for antidepressants (Allen, 2000). In response to these regulations, the use of antipsychotics, antianxiety drugs, and hypnotics in nursing homes decreased in the early 1990s. Since at least 1995, however, the use of antipsychotics in particular has been increasing. By 2001, antipsychotic use in nursing homes was approaching preregulation levels, was inappropriate (according to outside regulatory guidelines) in the majority of cases, and was of questionable effectiveness (Briesacher et al., 2005).

The risks and dangers of antianxiety drugs and hypnotics are well known, with antipsychotic drugs perhaps presenting even greater risks. In 2005, the FDA issued a public health advisory (U.S. Food and Drug Administration, 2005) warning that atypical antipsychotics (e.g., Risperdal, Zyprexa, Seroquel, Clozaril, Abilify, Geodon) are associated with increased mortality when used to treat behavioral problems in older people with dementia (this is the most common reason antipsychotics are used in nursing homes). The FDA now asks that the labeling for these drugs include a warning about this risk of death and a statement that these drugs are not approved for such use. Since the FDA advisory, further evidence suggests that conventional antipsychotics (e.g., Haldol) have similar risks (Wang et al., 2005). It remains to be seen whether the FDA warning will decrease the use of antipsychotics.

Antidepressants have so far escaped such regulatory disapproval; indeed, their use is somewhat encouraged by various regulatory provisions. However, their long-term safety has not been established. Taking them for years may or may not have significant risk. The risks of many psychotropic drugs have not become apparent until they have been used for many years, and it is possible that problems with antidepressants may yet emerge (Glenmullen, 2000; Valenstein, 1998).

All of these informal considerations relevant to the referral process suggest the importance of efforts to ensure that physicians and other long-term care staff who make referrals understand the value and potential benefits of psychological services.

Types of Referrals

In reviewing the types of referrals typically made to psychologists in long-term care settings, it is important to note that patients are often referred for multiple reasons. The problems on the list that follows are not mutually exclusive and frequently coexist. Patients with dementia can have adjustment disorders with feelings of depression, for example. Patients with personality disorders can have anxiety problems. Above all, many patients in long-term care settings, perhaps most, have significant medical problems along with whatever psychological problems they may have. The interaction between the two adds a level of complexity not present in medically healthy patients. With any referral, always consider that medical issues may be a cause or an effect (or both) of psychological problems a patient is experiencing.

Dementia and Other Cognitive Disorders

Patients with dementia or other cognitive disorders may be referred for a variety of reasons. A patient with apparent cognitive impairment may be referred for differential diagnosis: Is the patient cognitively impaired, and if so, why, to what degree, and in what way? Patients in long-term care often are referred with a diagnosis of dementia, which may or may not be accurate. On occasion, patients are admitted to a nursing home from a hospital with a diagnosis of dementia even though they have no cognitive impairment at all. Typically, they had a delirium in the hospital, which was misdiagnosed as dementia, and is now resolved. Patients who do have cognitive impairment may have other conditions instead of or in addition to dementia. Delirium, perhaps due to some acute underlying medical or physiological condition or adverse reactions to drugs, may be present. The history may suggest or reveal that a patient has mental retardation. The patient may have focal cognitive or executive function impairment(s) from a stroke along with areas of intact function, rather than the more diffuse impairment of dementia. Among patients who do have dementia, the question may be whether the etiology is probably Alzheimer's disease, vascular disease, or some other dementing disorder. Answering such differential diagnosis questions may inform patients' medical treatment and how staff understand and manage issues that arise.

Patients with dementia and other cognitive disorders are also often referred because of behavioral problems. Patients with Alzheimer's or vascular dementia may wander, be agitated, restless, or combative, or display other problem behaviors.

More subtle problems may be presented by patients who have had a stroke (typically a right-hemisphere stroke), who seem relatively intact, and yet may be disinhibited and interpersonally insensitive and difficult. Responding to these referrals often involves helping staff understand the behaviors at issue, giving staff input into care planning, and suggesting guidelines or protocols for behavior management.

Depression

Patients in long-term care may be referred for any of a broad spectrum of depressive disorders, including both "major" and "minor" depressions. The incidence of depression in nursing homes, in particular, is thought to be high. Factors contributing to the depression may be more or less apparent, although in long-term care there is often no shortage of reasons why a patient might be depressed. Patients may be depressed over serious medical illness, with significant pain or disability. They may have difficulty adjusting to placement in a nursing home, or be mourning the loss of their home and previous life. Their depression may focus on the loss of a spouse, the deaths of their peers, or other untoward life events. In long-term care settings, the events that precipitate and maintain depression are often undeniable losses that would be upsetting and challenging for anyone. They are facts that cannot be changed by anything the patient does, but that must simply be adapted to or coped with in some way.

Anxiety

The same kinds of events that precipitate depressive disorders in long-term care settings sometimes trigger anxiety instead of or in addition to depression. People with a variety of anxiety disorders may be referred for treatment. Patients who have chronic obstructive pulmonary disease (COPD) and thus suffer from chronic breathing difficulties, for example, are often very anxious and are often afraid to leave their oxygen-equipped rooms for even a moment for fear that they will suffocate. Patients who have fallen and broken a hip, to cite another example, are sometimes extremely anxious about trying to stand and walk again and may become disabled and wheelchair-bound, if not helped to overcome their fear.

Behavior Problems

Referrals for behavior problems are frequent in long-term care settings and are usually made when the patient is being uncooperative or noncompliant in some way. The problem may be a behavioral "excess;" wandering, getting out of bed and falling, calling for staff repeatedly, and cursing at or insulting staff are among many examples. Or, the problem may be a behavioral "deficit," where the resident is not engaging in some desired behavior; refusing to eat or take medication or refusing to get out of bed, for example.

Behavior problems can arise for any number of reasons, of course. Dementia, depression, anxiety, personality factors, or many other conditions may underlie the behavior(s). Uncovering the problem's etiology and communicating it to the physician and other staff will greatly affect how the problem is handled. A patient might refuse medication because he or she has dementia and cannot understand its purpose and value; the patient might be depressed and not care about the consequences of refusing the medication; or the refusal might be an informed, considered one, where the patient is rationally exercising his or her right to refuse medical treatment.

Dementia, as noted earlier, is a frequent cause of behavior problems, and often leads to referrals asking for guidance or suggestions for management. Behavior problems also frequently stem from personality disorders. Perhaps not surprisingly, patients with such disorders seem to be at high risk for having difficulty adjusting to living in a long-term care setting and for developing conflicts with staff. They are often perceived by staff as extremely demanding, uncooperative, argumentative, unreasonable, or abusive, for example, and often arouse correspondingly strong and sometimes angry and judgmental reactions in staff. Responding to such referrals can be very challenging and usually involves working with both patient and staff to resolve or manage the problem.

Perhaps more than any other type of referral, behavior problems raise difficult professional questions for the psychologist. "Problems" are in the eye of the beholder. Long-term care settings and the providers who staff them have rules, regulations, routines, responsibilities, and expectations that are generally but not always reasonable. Patients, on the other hand, have rights to autonomy and self-determination, and expectations and desires of their own, sometimes reasonable, sometimes not. Disagreements inevitably arise from time to time; staff says the patient's behavior is the problem, and the patient says the staff's behavior is the problem. Who should change and where do the psychologist's professional and ethical responsibilities and duties lie? A cognitively intact patient, for example, might assert her right to get in and out of bed without assistance whenever she wants even while acknowledging some risk of falling. Staff might see the risk of falling as unacceptably high and see the patient's refusal to always call and wait for assistance as a problem behavior. In principle, the psychologist's first duty must be to the patient and his or her well-being, not to an institution, but translating this principle into helpful action in specific cases, and reconciling the patient's and the institution's perspectives when possible, often calls for very clear thinking, diplomatic skill, and good judgment, both practical and ethical.

Medical Illness

Problems associated with medical illness frequently provide reason for referral, and as noted earlier, the relationship between psychological problems and medical illness can be one of both cause and effect. The full range of medical illnesses is present in long-term care settings, and commonly includes stroke, heart disease, cancer, orthopedic conditions such as hip fracture and hip replacement, diabetes with all its complications, including amputation, neurological and neuromuscular disorders such as Parkinson's disease and multiple sclerosis, and pulmonary disorders, among many others. Patients with such illnesses often react with depression, anxiety, anger, or other emotions, and concern about their medical condition often dominates their thoughts and feelings. In addition, illnesses that affect brain function, stroke, or cancer that metastasizes to the brain, for example, can directly (i.e., neurologically) affect patients' cognitive function and behavior, causing cognitive impairments, emotional lability, impulse control problems, impaired judgment, and other problems.

Medical illness can be the effect of psychological problems as well, the extreme case of course being psychosomatic illness. But, even aside from this extreme, psychological problems can worsen or exacerbate already-existing medical illness. Patients with COPD who are highly anxious can make their breathing difficulties even worse; patients with diabetes who become depressed and fail to eat or otherwise care for themselves properly can worsen their condition and make complications much more likely. Examples can be multiplied indefinitely.

Patients with terminal illness are a special subset of patients with medical illness, and are sometimes referred for psychological services. For such patients, a focus on palliative care is increasingly recognized by most healthcare disciplines as the appropriate treatment approach. For psychology, such an approach might involve focusing on easing the patient's emotional distress as much as possible, helping to clarify and communicate the patient's directives about his or her care, and addressing issues of death and dying and other existential concerns, or not addressing them, as the patient wishes.

Rehabilitation

Many nursing home patients are admitted for rehabilitation, to recover from the effects of a recent episode of medical illness as much as possible, and then hopefully, return to the community. The success of this plan depends in part on the patient's understanding of, engagement in, and motivation for rehabilitation. This is true in assisted-living communities as well, where many new residents are recently discharged from a hospital or rehabilitation facility and continuing physical therapy or other rehabilitation is expected. When psychological problems appear to be interfering with the patient's progress, a referral for psychological services may be made. Since the time allowed for rehabilitation is strictly limited and progress is tightly monitored by almost all insurers, an early and prompt referral, at the first sign of any problem, is crucial. If the referral is delayed, it may come too late, as the patient's progress may have already stalled and coverage may be terminated by the insurer before any psychological intervention has time to take effect.

Other Focused Problems

Some referrals come with a relatively focused referral question. One common question is whether the patient has the capacity to make healthcare, financial, or other decisions. The question of capacity for healthcare decisions, in particular, may sometimes be urgent; the need for a decision about some treatment or procedure might be immediate, and the patient's physician might want prompt advice about how to proceed. Can the patient make an informed refusal of a medication or treatment? Conversely, can the patient give informed consent for a surgical procedure?

Another common referral question is suicide risk. Typically, this is an urgent referral made after staff has reported some statement by the patient that may indicate a wish to die or an intention to harm him or herself. This question is usually taken very seriously in long-term care facilities and should generally be answered as if it were an emergency. Is there any substantial risk? If so, does the patient require hospitalization or can he or she be managed by staff with one-to-one coverage or other measures?

In Closing

This brief review of the types of referrals for psychological services that are typically made in long-term care settings hopefully conveys the interest, variety, challenge, and rewards that psychologists working in these settings enjoy.

Summary

The formal referral process is complex, varying from case to case and setting to setting, because different insurance carriers have different requirements and procedures and because state and federal laws and regulations set additional requirements. The informal aspects of the referral process (e.g., who initiates the referral) also vary, depending on the nature of the facility involved. The types of referrals typically made to psychologists include patients with dementia and other cognitive disorders, depression, anxiety, behavior problems, medical illness, rehabilitation issues, questions of decision-making capacity, or questions of suicide risk.

References

Allen, J. E.(2000). *Nursing home federal requirements and guidelines to surveyors*(4th ed.). New York: Springer.

Briesacher, B. A., Limcangco, M. R., Simoni-Wastila, L., et al.(2005). The quality of antipsychotic drug prescribing in nursing homes. *Archives of Internal Medicine*, 165, 1280–1285.

Glenmullen, J.(2000). *Prozac backlash: Overcoming the dangers of Prozac, Zoloft, Paxil, and
other antidepressants with safe effective alternatives.* New York: Simon and Schuster.

U.S. Food and Drug Administration. (2005). FDA public health advisory: Deaths with antipsychotics in elderly patients with behavioral disturbances. Retrieved May 3, 2005, from http://
www.fda.gov/cder/drug/advisory/antipsychotics.htm

Valenstein, E. S.(1998). Blaming the brain: The truth about drugs and mental health. New York:
The Free Press.

Wang, P. S., Schneeweiss, S., Avorn, J., et al.(2005). Risk of death in elderly users of conventional
vs. atypical antipsychotic medications. *The New England Journal of Medicine*, 353, 2335–2341.

Chapter 3
Assessment

Barry A. Edelstein, Lynn E. Northrop, and Lisa M. MacDonald

The assessment of older adults in long-term care is always challenging, often perplexing, and occasionally frustrating. Physical and mental health problems, and the positive and adverse effects of multiple medications, are often the principal sources of assessment complexity. The clinician must also consider these factors in the context of age-related changes in biological, psychological, and adaptive functioning. Finally, the client who is to be assessed is not always available or cooperative, and may require multiple assessment sessions because of factors including mood, illness, pain, and fatigue.

Responsible Assessment of Older Adults

Use Psychometrically Sound Assessment Instruments. A wide range of assessment instruments have been used to assess psychopathology in older adults, but several lack psychometric support for use with older adults in general, and long-term care residents in particular. Two of the more important forms of validity for this population are content and construct validity. Content validity is an index of the extent to which an instrument samples the domain of interest. Experiences and symptoms of older adults often differ from those of younger adults, raising questions about the content validity of instruments developed for younger adults when used with older adults. For example, Kogan and Edelstein (2004) found evidence that younger and older adults experience different fears and that traditional measures of fear do not include many of the most frequently reported fears of older adults. Construct validity, an index of how well the instrument measures the construct of interest, must be demonstrated with older adults, as age-related changes in psychological constructs have been demonstrated (e.g., Kaszniak, 1990).

Use Age "Norms" When Available. Most of the commonly used assessment instruments are norm-based. Many offer no normative data for older adults, and even fewer offer norms for long-term care residents. One must therefore be cautious when interpreting scores of norm-based instruments if older adults are not included in the normative group.

E. Rosowsky et al. (eds.), *Geropsychology and Long Term Care*,
DOI: 10.1007/978-0-387-72648-9_3, © Springer Science+Business Media, LLC 2009

Conduct Culturally Competent Assessments. An accurate and meaningful assessment requires consideration of the cultural and ethnic background of the individual. Factors to consider include the degree to which an individual identifies with and is involved in a particular culture, unique culture-specific psychological disorders (e.g., "nervios" or "susto" in Latino populations) and cultural influences on the presentation of psychological disorders (Edelstein, Martin, & Koven, 2003). Without an adequate understanding of how culture influences the expression of psychological symptoms, inaccurate diagnosis may be made and proper treatment may not be offered.

An additional concern is the use of assessment measures with older adult minority individuals when those measures were created for and normed on Caucasian individuals. There is a growing body of literature regarding the use of specific assessment measures for use with individuals from various cultures. The reader is encouraged to become familiar with this literature when assessing an older adult from an ethnic or cultural minority group (e.g., Baird, Ford, & Podell, 2007; Baker & Espino, 1997; Cheng & Chan, 2004; Lee et al., 2004; Lucas et al., 2005; Mui & Shibusawa, 2003; Nuevo, Mackintosh, Gatz, Montorio, & Wetherell, 2007; Yamada, Valle, Barrio, & Jeste, 2006). Although significantly more effort is needed to develop culturally appropriate measures for use with minority older adults, and in validating translated versions of English language measures into other languages, there is preliminary support for the use of some translated assessment measures (see Novy, Stanley, Averill, & Daza, 2001).

Adjust Methods to Accommodate for Physical and Sensory Limitations. Knowledge of age-related sensory, cognitive, and motor skills (see Whitbourne, 1999) and the potential effects of deficits in these domains on assessment process and outcomes are very important. In general, the clinician should initially determine whether the resident has visual or other perceptual difficulties. The clinician then should take steps to minimize the impact on assessment. For example, if the resident has an eyeglass prescription, be sure he or she is wearing the eyeglasses. To minimize the effects of cataracts, one can avoid using high-gloss paper for self-report inventories, visual aids, and figure drawing tasks. Also, one should avoid having the client face a window during the daytime, which can contribute glare. Be sure that the setting, including your face, is adequately lighted. Sit close enough to be clearly seen, and avoid the situation in which you are seated with a window behind you, which can throw a shadow on your face and increase the likelihood of glare on the assessment materials. The light should be bright, but not glaring. For residents with severe visual deficits, one can still be adaptive in one's approach. For example, one could limit nonverbal directions, relying more heavily upon verbal or kinesthetic cueing. One can also use multimodal (say and do) directions when possible. Various diseases (e.g., diabetes mellitus, cataracts, glaucoma, macular degeneration, myotonic dystrophy, hypoparathyroidism, Wilson's disease) and medications (e.g., phenothiazines, corticosteroids, antibiotics, and antimalarials) can also affect the visual system.

Hearing loss can also compromise assessment. It is important to ask the resident if he or she has a hearing aid and, if so, encourage the resident to use it during your session. A quiet environment, with few auditory distractions, can minimize some of the drawbacks of hearing aid use. This also may enhance hearing directly by minimizing some of the "excess amplification" that sometimes occurs with the use of hearing aids. Some residents will compensate for hearing loss by reading lips. Consequently, it is important to not overarticulate, which can distort speech and facial gestures. Lip reading is a stressful and demanding task, so breaks should be considered during evaluations and therapy sessions. If assessing a hearing impaired resident, one might consider providing the test questions in writing. Assistive devices such as audio listening aids and amplifiers can be quite helpful. Edelstein, Martin, and Goodie (2000) offer a more detailed discussion of sensory, motor, and cognitive deficits and methods for minimizing their effects.

Recognize That Older Adults May Not Be Familiar with Testing. The manner in which assessments are conducted with older adults influences the value and utility of the information gathered. The current cohort of older adults is not as familiar with the process of testing and evaluation as younger adults, and standardized assessments are often experienced as intimidating, threatening, or irrelevant (Hays, 1996). Older adults, particularly those who are medically fragile or who are referred for mental health problems, can be more susceptible to fatigue, or poor attention and concentration than younger adults are. Therefore, some assessment measures can be overwhelming because of their length, or intimidating because of the content of the items (e.g. MMPI-2).

Consider the Medical and Pharmacological Factors. Many common physical disorders experienced by older adults have psychological symptoms. In fact, some physical disorders are mistaken for psychological disorders and go undetected (see Frazer, Leicht, & Baker, 1996). Thus, it is important for clinicians to learn the psychological concomitants and manifestations of chronic diseases (see Frazer et al.). One might also become accustomed to checking the results of laboratory testing when they are accessible, as they can reveal untreated or inadequately treated disorders or conditions that can have psychological presentations (e.g., hyper- or hypothyroidism). Many medications, both prescription and over-the-counter, can cause psychiatric symptoms in older adults, even those medications that are well tolerated in younger adults (e.g. Benadryl). In short, clinicians assessing older adults must be prepared to rule out or take into consideration the effects of aging, diseases, and medications before appealing to the environment and psychological variables for answers.

Use Appropriate Methods for Cognitively-Impaired Individuals. As one moves from cognitively intact to cognitively impaired individuals, there is a necessary shift from more traditional to more behavioral, idiographic approaches to assessment. Moderate to severe cognitive impairment typically precludes accurate and reliable self-report. Thus, assessment is less likely to focus on the person's personality, cognitions, and self-reported behavior, and more likely to focus on the person's observed behavior and the environmental conditions that could be maintaining it. The question, "What is it about this person that causes him to behave this way" becomes, "What is the function of this behavior and under what conditions is this person exhibiting this behavior?" Questions asked might include, "What time of day does the problem occur?;" "In whose presence does it occur?;" "How often does the behavior occur?;" and "What happens after the behavior occurs?" Of equal importance is the question of the conditions under which the behavior does *not* occur. The assessment methods become more circumscribed and direct, relying principally upon report by others and direct observation.

Consider the Unique Presentation of Axis I Disorders Among Older Adults. Older adults may be less likely to report depressed mood or emotional problems. In contrast to younger adults, the presentation of depression among older adults is more likely to include irritability, worry, somatic complaints, cognitive impairment, deterioration of self-care, social withdrawal, and paranoia (Sable, Dunn, & Zisook, 2002). Kasl-Godley, Gatz, and Fiske (1998) note that loss of interest, lack of energy, and sleep disturbance can be useful distinguishing features between depressed and nondepressed older adults. They also note that delusions may be more common in older than younger depressed adults. In a large multisite, multicohort study with over 1,400 subjects, Husain et al. (2005) found that older patients endorsed longer durations of illness, more major depressive episodes, a later age at onset of their first major depressive episode, and more general medical comorbidities. They had more middle and terminal insomnia, less irritability, and less hypersomnia. They were less likely to hold negative views of themselves or of their future and were less likely to report previous suicide attempts.

Assessment of anxiety may be complicated by comorbid medical illness and treatment, somatic symptom presentation, and the reluctance of some older adults to acknowledge psychological difficulties (Blazer, George, & Hughes, 1991). The content of the worrisome thoughts changes over the life span; for example, older adults are more likely to worry about health-related problems (Scogin, 1998). Thus, despite high rates of anxiety symptoms and disorders among older adults, anxiety often goes unrecognized and untreated in this population (Wetherell, Lenze, & Stanley, 2005).

Clarify the Referral Question. A critical first step in conducting an assessment is having a clear understanding of the referral question. As discussed in Chap. 2, in skilled nursing facilities (SNFs) and in assisted living (AL), requests for assessments come from a variety of sources, including physicians, nurses, social workers, physical therapists, activities staff, admissions workers, and even administrators. In an effort to clarify referrals, clinicians should seek answers to questions such as

- What triggered the referral? Was the referral triggered by a change in the resident's status? A family concern? An incident of inappropriate behavior?
- Who will be the "consumers" of the assessment results? Will the consumers be facility staff members, or is it likely that the report will be read by family members or go outside the facility?
- How do the consumers hope to utilize the information that you provide? Are they simply looking for assistance in making an accurate diagnosis, or will your results be used in the interdisciplinary care planning process, discharge planning, or in determination of a resident's capacity to participate in medical or financial decision-making?

Answers to these questions will help guide the clinician in deciding what additional psychological tests or other assessment tools and methods are needed, as well as the content of the consultation report, treatment plan, and recommendations.

Clinicians should be prepared for the possibility that the referral question will reflect the lack of psychological savvy of the referrer as much or more than it informs the clinician about the resident's needs. Many referring parties have had little or no mental health training and have never used or requested psychological services. They may be unaware of the differences between a psychologist and a psychiatrist, or between a psychologist and a clinical social worker. They may not understand the breadth and depth of information that can be provided with a psychological evaluation. Thus, in the process of helping the referrer clarify the reason for referral, one can expand that person's understanding of the behavioral and psychological problems assessed and addressed by psychologists.

The Internet offers a variety of free Web sites that provide updates on relevant topics. For example, the American Psychological Association's Office on Aging Web site offers a document on capacity assessment. While intended for lawyers, it is very useful for clinicians. Go to http://www.apa.org/pi/aging/ to get it and other useful information regarding work with older adults.

Methods of Assessment

When selecting the methods of assessment for use with an older adult, the clinician must consider the goals of the assessment, the length of time for assessment, the availability of other informants, the stamina of the older person, and the availability of appropriate norms for the age, gender, education, and ethnicity of the patient.

Each method or source of information has limitations, for a variety of reasons, which is why multiple methods (e.g., self-report, report by others, direct observation) are encouraged. For example, one might measure depression of a nursing home resident by using a combination of two or more of the following: a self-report instrument, a rating scale completed by a staff member, direct observation of relevant behavior, and a brief clinical interview completed by a mental health professional.

Clinical Interview, Mental Status Examination, and History. The most common source of clinical data is the clinical interview, which is the most flexible and convenient assessment method. One must be mindful that the validity, reliability, and accuracy of self-report information diminish as cognitive functioning declines. Moreover, older adults are more likely than younger adults to be more cautious when responding, give more acquiescent responses, refuse to answer certain types of questions, and to respond "don't know" (Edelstein et al., 2003). For residents whose cognitive skills permit, a thorough history and mental status examination is a first step toward a thorough assessment. Having a resident describe his or her history is a good method for establishing rapport, a potentially good source of information regarding immediate and remote memory, and often a good way of beginning an interview. Though much of the history sometimes can be gleaned from the resident's chart, a resident's historical account can yield additional information and provide the clinician opportunities to question the resident regarding issues and information revealed. In addition, discrepancies between information provided by a resident and that available in the medical record may alert the clinician to problems with or fluctuations in the resident's cognitive functioning. This, in itself, is valuable information to report and comment on in the assessment results.

Screening Instruments. Screening instruments for a variety of problem areas (e.g., cognitive functioning, depression) are available. Such instruments offer an expedient approach to cognitive assessment, which can be followed with more in-depth assessment as indicated. One must be aware of the limitations of such screening devices, however. For example, cognitive screening instruments are not sensitive to all potential cognitive deficits. Screening instruments for depression rarely cover all of the criteria needed for a diagnosis of depression. The use of "screening" instruments poses special problems if the clinician hopes to be reimbursed by Medicare. Typically, Local Coverage Determination Policies (LCDs) state that mental health and cognitive screening are not authorized (i.e., reimbursable) services. LCDs hold that there must be an existing "medical necessity" before a service can be provided. Therefore, if the clinician suspects a psychiatric or cognitive disturbance and then performs an assessment, the LCD would not consider this a screen. However, if, for example, the clinician were to perform a mental health or cognitive screen on all new admissions to identify people in need of further assessment or intervention, Medicare would likely disallow this service. Such services would have to be performed gratis or billed directly to the facility or family.

In choosing an assessment tool, it is important to consider both the sensitivity and specificity of the various measures, and to select tests based on the goals of the assessment. For screening, a test with high sensitivity is preferred; however, a test with high sensitivity will yield both a high rate of true positives as well as false positives, and the examiner must then conduct further evaluation. Screening measures should also have at least moderate (60%) specificity to limit the rate of false positives. After detection of possible dysfunction, a test with high specificity is preferred to assist in further evaluating the cause of the problems. Clinicians can limit the rate of false positives by using multiple data sources, such as combining a brief screening for dementia instrument with an informant report interview, such as the Informant Questionnaire on Cognitive Decline in the Elderly (IQCODE; Jorm & Jacomb, 1989). This instrument is available at http://www.anu.edu.au/iqcode/, including short form, long form, and retrospective form, in several languages. It asks the informant to compare his or her friend or loved one's current cognitive functioning to their functioning ten years ago. Use of multiple methods of assessment can rule out false positives, preventing the needless time, expense, and demands on the patient of conducting a full cognitive assessment (Jorm, 2004; Scanlan & Borson, 2001).

Self-Report and Questionnaires. Though self-report is the most common assessment method, the evidence supporting the accuracy, reliability, and validity of older adult self-reports is mixed, as noted above. For example, older adult estimates of their functional ability have been questioned, with some overestimating their functional ability and others both under- and overestimating their abilities. Similarly, self-reports of memory impairment among older adults may be inaccurate. In contrast, older adult self-reports of insomnia have been found to be reasonably accurate. Though some self-report instruments may become invalid when used with cognitively impaired residents, others may be acceptable. For example, self-report instruments designed to measure mood may be utilized with older adults experiencing mild-to-moderate dementia, as accurate self-report of recent mood requires only minimal memory ability. The use of multiple assessment methods can offset some of the limitations of self-report. Appropriately, most Medicare carriers do not allow self-report assessments to be billed as psychological testing, unless scoring and interpretation require sophisticated psychological expertise (e.g., MMPI). This remains true even if the self-report measure is administered as an interview to compensate for the older adult's sensory and/or cognitive deficit(s). Nonetheless, self-report can often reveal the patient's own perception of his or her functioning, whether supported by objective results or not, thereby potentially offering a valuable insight into the patient's own interpretation of his or her condition.

An Interdisciplinary Approach. Reports by other caregivers can be very important sources of information, as noted in the previous sections of this chapter and text. They can be used to establish a medical and psychological history that could inform one's current evaluation, substantiate the reports of residents, and can be essential when residents are incapable of reporting reliable and valid information.

In SNFs and ALFs, certified nursing assistants (CNAs) and other noncertified aide staff are required to be present around the clock; though in AL settings, staff to resident ratios are certainly lower. Aides spend virtually all of their shifts in direct contact with residents – more than perhaps any other discipline. In the process of providing personal care, or simply because of frequent informal contacts, they often develop close relationships with residents. Aides' observations of residents can be considered in a thorough assessment, though the objectivity and training of the staff member must also be taken into account. With administrative support, aides can be trained to formally observe and document specific resident behaviors.

Direct Observation of Behavior. Direct observation of behavior can be one of the richest and most accurate assessment methods. Behavior observation can begin the moment one sees a resident in the hall, or waiting area, and can continue throughout the clinical interview and formal testing. It is important to note that unless formally trained in direct observation and assessment, staff members are not always the best observers or informants regarding resident behavior. The importance of direct observation by a trained clinician cannot be overemphasized. Staff members may be biased in what is reported, and frequently are not trained to identify or record behavior or environmental events that the clinician might deem useful in the assessment process. There are several advantages of using direct observation, particularly when assessing older adults who are uncooperative, unavailable for self-report, or suffer severe cognitive or physical impairment. In addition, simple observational procedures can be easily taught to people with little or no previous experience. Unfortunately, at the present time, Medicare will not reimburse nonphysician providers for time spent observing behavior. Just recently, however, case management procedure codes have been added to the list of services reimbursed by Medicare, as long as the psychologist consults with two or more additional disciplines during that consultation session.

Multiple Informants to Involve in Assessment

Multiple sources of information about an older adult are valuable and often essential to the completion of a thorough assessment. While it is true that the older adult is perhaps the most important informant, it is also true that virtually everyone who has observed the older adult is a potential and often unique source of information. In long-term care settings, there are many individuals who can offer unique perspectives on the residents and whose information complements our assessment.

Close Friends and Family Members. Close friends and family members are generally good sources of historical information and may be the best resource for understanding the factors that led to placement, and for providing a sense of what level of functioning the individual might be capable of achieving with effective intervention and how the resident is progressing. They can also alert the clinician to early signs of any decline in functioning, if any.

It should be noted that, appropriate releases of information must be in place before contacting friends and family for information, or otherwise involving them in an individual's care, whether obtained from the resident, or in the absence of decision-making capacity, from the responsible party or conservator.

Dieticians and Food Service Workers. Anyone involved in meal times (food servers, nurses aids, kitchen staff, etc.) observe important resident behaviors. A registered dietician (RD) makes recommendations on diet relative to specific medical issues (e.g., diabetes, hypertension), and chewing or swallowing problems, and can be excellent allies in determining why a resident might have eating difficulties.

PCPs – Physicians and Nurse Practitioners. Medicare requires psychologists to attempt consultation with patients' primary care provider (PCP), regardless of where the patient resides. Conferring with the PCP is highly desirable, and both will gain in their knowledge of and approach to the patient from this collaboration.

Registered and Licensed Nurses. Both SNFs and ALFs and some board and care facilities employ nurses who can be allies in formal assessment and routinely assess and document physical status (i.e., pain, sleep, and weight), mood, specific behaviors, mental status, and other factors that both improve and interfere with caregiving activities.

Pharmacists. Pharmacists are an invaluable source of information regarding the desirable and adverse effects of medications and the potential consequences of drug interactions. They often can offer insights on how medications may be creating paradoxical reactions instead of the desired effects.

Recreational Therapists and Activities Professionals. Recreational therapists and activities professionals can provide information about the older adult's preferences for activities and interactions with other residents or activities staff. Such information can also supplement the psychological assessment of the patient.

Rehabilitation Staff. Physical, occupational, and speech therapists work closely with residents in their rehabilitation. Consequently their assessments yield valuable information on motivation, outlook, and acceptance of their limitations, as well as what therapeutic style and approaches elicit the best response from the patient.

Social Services Staff. Social services personnel are very important sources of information about residents, and are often critical allies in the development and maintenance of a viable, comprehensive mental health program. Their assessments and care plan approaches incorporate key factors relevant to residents' psychological, interpersonal, behavioral, and family functioning. This information is a rich source of data about the resident.

Multiple Dimensions of Assessment

"Healthcare and social service providers and organizations tend to specialize, but human beings are general entities with multidimensional functions, needs, and problems" (Janik & Wells, 1982, p. 45). For most humans, advanced age and the need for long-term care go hand-in-hand with a clinical complexity of many layers. Residents in long-term care, particularly in SNFs, are likely to have multiple medical problems, take several different medications, and have one or more deficits in their ability to perform activities (and instrumental activities) of daily living (i.e., ADLs and IADLs, respectively). Mental health problems further complicate the clinical picture.

The complexity of performing assessments and providing care can be increased through the involvement of extended family systems that may or may not be involved in the resident's care (i.e., spouses, adult children, grandchildren). Nursing facilities can add additional layers of complexity through their diverse networks of care staff, and flurries of activities, sights, sounds, and smells. Competent assessment in this setting requires an understanding of how each of these many elements comes into play for the individual resident.

Medical: Physical Health and Medications. Accurate assessment of physical health is often made more difficult by the interplay between illnesses and the medications used to treat them. Sometimes prescribed medications (individually or in combination) cause symptoms that might themselves be misinterpreted as symptoms of additional physical or psychiatric disorders. For example, some of the inhalants used to treat COPD have the side effect of anxiety. Hallucinations, illusions, insomnia, and psychotic symptoms are possible side effects of various antiparkinsonian agents (Salzman, 2005).

Teasing out the individual and overlapping effects of psychiatric and physical illnesses is challenging. Physical illnesses can manifest with emotional or behavioral symptoms, and psychiatric illnesses sometimes present with physical symptoms (Morrison, 1997). For example, a seemingly simple medical problem such as a urinary tract infection (UTI) can cause delirium in an older adult, and thus can lead to impaired cognition, agitation, and even hallucinations. If the UTI goes unidentified, the older adult might inappropriately be labeled with a psychiatric diagnosis or characterized as having a behavior problem, possibly leading to inappropriate treatment of the psychiatric symptoms and failure to treat the UTI. An uninformed psychologist may develop an environmental or behavioral management program for problem behaviors that actually stem from the UTI. Bacterial and viral infections in older adults should always be considered when changes in behavior occur.

Depression offers a prime example of how physical and psychiatric symptoms can be shared by various disorders. Hypothyroidism, rheumatoid arthritis, pancreatic cancer, Parkinson's disease, diabetes, and COPD, for example, all increase the likelihood of depressive symptoms. Some medical disorders can even first present as depression (e.g., pancreatic cancer, Parkinson's disease).

The assessment of physical functioning is typically conducted by a registered nurse or physician and includes both a physical examination and laboratory tests (e.g., thyroid function, blood sugar, vitamin B12, folic acid, lipids). It is important for the psychologist to become familiar with laboratory tests for disorders that can present with psychiatric symptoms (e.g., hypothyroidism, hyperthyroidism), and to communicate with medical staff regarding the physical health of the resident. Because the psychologist is asked to assess the patient when emotional and behavioral symptoms emerge, he or she must be aware of potential sources of these symptoms. For further reading on this topic see Morrison (1997).

Assessment of Functional Activities of Daily Living and Instrumental ADLs

Individuals requiring skilled nursing care may have deficits in their ability to perform activities of daily living (ADLs), such as maintaining hygiene, toileting, or eating. Equally important is a diminished ability to perform more complex tasks, such as preparing meals, paying bills, and driving. These more complex tasks are often referred to as instrumental ADLs or IADLs. Independent performance of ADLs and IADLs can be impaired by a variety of problems ranging from acute and chronic disease (e.g., viral infections, atherosclerosis, chronic obstructive pulmonary disease, diabetes) to various forms of psychopathology (e.g., depression, dementia, substance abuse, and schizophrenia; La Rue, 1992). Normal age-related changes can also diminish one's level of adaptive functioning (see Edelstein, Goodie, & Martin, 2004). For example, age-related loss of bone density and muscle strength can limit a wide range of ADLs (e.g., walking, house cleaning).

ADLs and IADLs can be assessed through self-report, direct observation, and report by others using standardized assessment instruments (e.g., Adult Functional Adaptive Behavior Scale, Spirrison & Pierce, 1992; Direct Assessment of Functional Status Scale, Lowenstein et al., 1989; Katz Activities of Daily Living Scale, Katz, Downs, Cash, & Gratz, 1970).

Assessment of Cognitive Functioning

The assessment of cognitive deficits is one of the more complex and important tasks in multidimensional assessment. Normal, age-related cognitive deficits must be distinguished from deficits due to a plethora of possible organic, psychiatric, and environmental causes. Age-related changes in cognitive functioning are not uncommon among older adults. However, these changes are typically observed only within certain domains (e.g., working memory), whereas other domains may evidence stability or even improvement (e.g., semantic memory) (Babcock & Salthouse, 1990; Light, 1992).

Diminished cognitive functioning may result from a variety of factors beyond "aging" (e.g., vitamin deficiency, drug side effects, cardiovascular disease, schizophrenia, brain injury, dementias). Accurate identification of reversible vs. irreversible causes of cognitive impairment is essential if proper interventions are to be employed. Similarly, the psychologist can play an important role in "setting the record straight" when a patient has received an inaccurate diagnosis. Unfortunately, benign and/or transient symptoms of cognitive impairment sometimes result in the patient being mislabeled with a diagnosis of dementia, for example. This label, though inaccurate, can influence the manner in which staff perceives and interacts with the patient, and can be very upsetting to the patient and his or her family. In addition, diagnoses that suggest cognitive impairment can influence Medicare and other payer decisions to reimburse for psychological treatment. Although Medicare cannot legally deny coverage simply because of a dementia diagnosis, the diagnosis can be used to trigger a review of medical necessity and result in delayed or denied reimbursement.

The starting point for cognitive assessment is typically the administration of a screening instrument. These instruments vary in length, content, validity, and utility. Such instruments are sometimes used to identify individuals who might warrant more extensive neuropsychological assessment (Alexopoulos & Mattis, 1991). Even when more extensive assessments are not warranted or possible, briefer assessments might provide data that are useful in developing recommendations for staff or in helping the resident understand his or her cognitive strengths and weaknesses. Caution should be exercised. Albert (1994), MacNeill and Lichtenberg (1999), and Ruchinskas and Curyto (2003) provide comprehensive and detailed reviews of screening instruments for cognitive functioning among older adults and sample instruments are described below.

The Mini-Mental State Exam (MMSE; Folstein, Folstein, & McHugh, 1975) is a brief measure of cognitive functioning with good reliability and validity in the geriatric population. It is sensitive to moderate and severe impairment, but may be less sensitive to mild impairment and subcortical deficits (Ruchinskas & Curyto, 2003; Tombaugh & McIntyre, 1992).

For those seeking alternatives to the MMSE that are brief, more sensitive, and noncopyrighted, the Montreal Cognitive Assessment (MCA; Nasreddine et al., 2005) and the St. Louis University Mental Status Examination (SLUMS; Tariq, Tumosa, Chibnal, Perry, & Morley, 2006) should be considered. These are both sensitive to mild cognitive impairment and require about the same amount of time to administer as the MMSE. Copies of the MCA (in several languages), references, and instructions can be found at http://www.mocatest.org. Copies of the SLUMS can be found at http://medschool.slu.edu/agingsuccessfully/pdfsurveys/slumsexam_05.pdf.

The modified 3MS (Teng & Chui, 1987) has been shown to be a good screening instrument for identifying dementia and all levels of cognitive impairment (Bland & Newman, 2001). The 3MS has a very low false-negative rate (less than 1%), making it useful for detecting most cases of dementia (Patterson et al., 1999). The 3MS is affected by age and education, and normative data are available (see Tschanz et al., 2002 for normative data). The 3MS takes slightly longer to administer than the MMSE (Bland & Newman, 2001).

The Mini-Cog (Borson, Scanlan, Brush, Vitaliano, & Dokmak, 2000), a brief dementia screening instrument with high sensitivity and specificity for detecting memory impairment, requires roughly 3 min to administer (Borson et al.; Lorentz, Scanlan, & Borson, 2002). The mini-cog was developed as a culturally sensitive measure and therefore is relatively free of language, education, and cultural biases, and has been useful for detecting a range of cognitive decline from subclinical cognitive impairment to severe dementia (Lorentz et al.).

The Dementia Rating Scale (DRS; Mattis, 1988) is a more lengthy evaluation of cognitive impairment that is useful for predicting everyday functioning for older adults up to 95 years (Holtzer, Burright, & Donovick, 2001) and in distinguishing various types of dementia (Lukatela et al., 2000). Administration time ranges from 20 to 45 min (Ruchinskas & Curyto, 2003). This measure has adequate cross-cultural validation (Arnold, Cuellar, & Guzman, 1998; Hohl, Grundman, Salmon, Thomas, & Thal, 1999).

The Neurobehavioral Cognitive Status Examination (COGNISTAT) is a screening instrument used for detecting cognitive dysfunction (Ruchinskas & Curyto, 2003) and requires roughly 10–25 min to administer. This instrument has good reliability and validity, but further normative data are needed for use with older adults. Overall, the psychometric data available for the COGNISTAT are less consistently positive than for the DRS or MMSE (Ruchinskas & Curyto).

The Cognitive Abilities Screening Instrument (CASI; Teng et al., 1994) is a mental status screening instrument developed for use with geriatric populations and found to be a reliable, cross-cultural measure of cognitive impairment. Stratified norms are available, as the measure is sensitive to age and education effects (Borson et al., 2000; Teng et al.).

The Short Portable Mental Status Questionnaire (SPMSQ; Pfeiffer, 1975) was designed for the geriatric population. This measure is generally good at identifying individuals with moderate to severe cognitive impairment, but has limited ability to identify mild cognitive impairment and in correctly classifying individuals in terms of their ability to independently perform ADLs (Lezak, 1995). Other measures developed specifically for use with older adults include the Repeatable Battery for the Assessment of Neuropsychological Status (RBANS; Randolph, Tierney, Mohr, & Chase, 1998) and the Cambridge Examination for Mental Disorders of the Elderly (CAMCOG; Roth et al., 1986). See Ruchinskas and Curyto (2003) for a review.

Informant rating scales can be useful for detecting dysfunction as well as measuring change longitudinally. They can be used for patients unable to complete self-report measures and standardized testing and to provide convergent data regarding a patient's behavioral symptoms and functioning. Consideration must be given to the informant's personality, relationship with the patient, and capacity to give accurate information (Jorm, 1997). The Informant Questionnaire for Cognitive Decline in the Elderly (IQCODE; Jorm, Scott, Cullen, & MacKimmon, 1991) rates cognitive abilities and can be used to measure change over time. A shortened 16-item questionnaire has proven equally effective as the original 26-item measure. The short form takes roughly 10 to 15 min to administer and does not appear to be affected by age, education, or gender (see Lorentz et al., 2002, for review).

When briefer tests reveal possible cognitive impairment, one might wish to pursue more detailed neuropsychological assessment, to refer for more extensive medical evaluation, or both. A wide range of neuropsychological assessment batteries have been used to further investigate cognitive functioning. These range from relatively small batteries focusing on dementia (e.g., Consortium to Establish a Registry for Alzheimer's Disease Neuropsychological Battery, Morris et al., 1989; Washington University Battery, Storandt, Botwinick, Danziger, Berg, & Hughers, 1984) to very comprehensive neuropsychological batteries (e.g., Halstead-Reitan).

Assessment of Mood Disorders and Anxiety. In the general population, older adults experience lower rates of some psychological disorders (e.g., depression and anxiety) than do younger adults (Blazer, 1994; Wolfe, Morrow, & Fredrickson, 1996). For example, the 1-month prevalence rate for anxiety among community-dwelling older adults (65+ years) is 5.5%, in contrast to 7.3% for younger adults (Reiger et al., 1993). However, among older adults residing in long-term care facilities, the prevalence of mental illness is higher than it is in the general population. As noted earlier, some epidemiological studies suggest that as many as 65% of older adults in nursing homes warrant a DSM diagnosis (Burns et al., 1993).

Psychological assessment of older adults often begins with an unstructured interview addressing a wide range of possible psychopathology, followed by more focused assessment that addresses identified problem areas. A variety of standardized assessment instruments have been used to assess psychopathology in older adults, but few have adequate psychometric support for use with this population. On a more positive note, there is some evidence to support a few of these instruments originally developed for use with younger adults. Though more are needed, a few psychopathology assessment instruments have been developed specifically for use with older adults. The following are commonly used measures of psychological functioning.

The Beck Depression Inventory-II (BDI-II; Beck, Steer, & Brown, 1996) is a 21-item multiple choice self-rating scale, which measures symptoms of depression, including somatic symptoms. There is research to support the use of the original version of the BDI (Beck, Ward, Mendelson, Mock, & Erbaugh, 1961) with older adults (Gallagher, Breckenridge, Steinmetz, & Thompson, 1983; Hyer & Blount, 1984; Norris, Gallagher, Wilson, & Winograd, 1987). And, recent validation of the BDI-II indicates similar validity correlations between the BDI-II and GDS in use with older women (Jefferson, Powers, & Pope, 2000). Because of the tendency for somatic items to be endorsed more frequently in the geriatric population, there has been concern about the BDI's tendency to overestimate depression in the elderly. However, Olin, Schneider, Eaton, Zemansky, and Pollock's (1992) study of psychiatric outpatients did not find a correlation between age and BDI score, nor an overendorsement of somatic items in older patients. That being said, however, this concern might still be relevant for inpatients or the medically frail.

The Hamilton Depression Rating Scale (Hamilton, 1960) is a widely used instrument for measuring depression. Despite its tradition in the field, it has several significant structural flaws. Some authors have concluded that there are better measures for assessing depression (Bagby, Ryder, Schuller, & Marshall, 2004).

The Center for Epidemiological Studies – Depression Sale (CES-D; Radloff, 1977) is a measure of depressive symptoms with sensitivity and specificity from 75% to 93% and 73% to 87%, respectively (Watson & Pignone, 2003). There is evidence to support the reliability and validity of the CES-D with older, ethnically diverse individuals (Areán & Miranda, 1997; Mackinnon, McCallum, Andrews, & Anderson, 1998; Mui, Bernette, & Chen, 2001). This measure is not appropriate for assessing depression in significantly cognitively impaired individuals (Lewinsohn, Seeley, Allen, & Roberts, 1997).

The Geriatric Depression Scale (GDS; Yesavage et al., 1982–1983) is a 30-item self-report instrument designed to assess the severity of depressive symptoms in older adults. Questions are posed in "yes" and "no" format. The GDS does not contain any somatic items that might inflate the scores of older adults. The GDS is available in many languages (see http://www.stanford.edu/~yesavage/GDS.html). Evidence for the reliability and validity of the GDS has been established for a range of cognitively intact older adults (Lesher, 1986; Norris et al., 1987; Rapp, Parisi, Walsh, & Wallace, 1988). The GDS has good sensitivity and specificity (Watson & Pignone, 2003); however, cognitive impairment affects the internal consistently of the GDS, and therefore, this measure is not appropriate for use with significantly cognitively impaired individuals.

To reduce the time necessary to administer the full 30-item GDS (Sheikh & Yesavage, 1986), many brief versions of the GDS have been developed beginning with the 15-item GDS-15 (Sheikh & Yesavage). The GDS-15 is able to detect presence and severity of depression (Isella, Villa, & Appollonio, 2001; Sheikh & Yesavage, 1986). The GSD-15 appears to be a useful tool for adults over 85 years of age, particularly if they have minimal cognitive dysfunction (de Craen, Heeren, & Gussekloo, 2003). More recently developed four- (GDS-4; Isella et al., 2001) and five-item (GDS – 5/15; Hoyl et al., 1999) GDS scales that show some potential as screens would be followed by more detailed investigation of depression, if indicated (Isella et al., 2001; Weeks, Mc Gann, Michaels, & Penninx, 2003). A short form of the GDS appears to have higher sensitivity and specificity than the full CES-D for nursing-home residents (Blank, Gruman, & Robison, 2004). A review of studies investigating the cross-cultural validity of the 15-item and 30-item GDS was inconclusive, and needs further exploration (Mui & Shibusawa, 2003; Stiles, 1998).

Müller-Thomsen, Arlt, Mann, Maß, and Ganzer (2005) reviewed the usefulness of instruments in detecting depression in Alzheimer's disease patients and found that the Montgomery and Asperg Depression Scale (MADRS; Montgomery & Asberg, 1979) and the Cornell Scale for Depression in Dementia (CSDD; Alexopoulos, Abrams, Young, & Shamoian, 1988) are useful with significantly cognitively impaired older adults (see Muller-Thomsen et al.). In particular, the CSDD may be the instrument of choice in the most severe cases of cognitive decline. Greenberg, Lantz, Likourezos, Burack, Chichin, and Carter (2004) found that while the CSDD underperformed in detecting depression in cases of mild-to-moderate cognitive impairment, it was the only measure to detect depression in patients with severe cognitive impairment. Sunderland et al. (1988) found only weak evidence for the concurrent validity of the Dementia Mood Assessment Scale (DMAS), and moderate interrater reliability estimates ($r = 0.74$ for core raters and $r = 0.69$ for other raters).

In addition to unipolar depression, bipolar disorder is certainly a reality for some older adults. As is true for the general population, there exist many more measures of depressive symptoms than of manic symptoms. To date there are no published measures of mania specifically for older adults. The Young Mania Rating Scale, which is a clinician-rated measure of manic symptoms, has been used with older adults (e.g., see Depp & Jeste, 2004). Readers should note that some research suggests early and late onset bipolar disorder may differ in etiology; hence, a thorough exploration of date of first manic episode is very important. For example, when mania has its first onset in late life, there is more likely to be a secondary cause such as vascular disease (Cassidy & Carrol, 2002), or other medical disorder or medication induced (Eastham, Jeste, & Young, 1998). This calls to mind the importance of considering medical and pharmacological issues in geriatric assessment.

There is a need for measures of anxiety developed specifically for older adults. Assessment of anxiety with scales developed for the general adult population is a concern when used with older adults because the presentation of anxiety in older adulthood tends to vary in severity and quality from that of younger adults (Scogin, 1998), and from the DSM-IV diagnostic criteria. Scales for assessment of anxiety specifically designed for older adults are needed, in part, because many of the symptoms of anxiety are also physical symptoms that might tend to occur with some regularity in older adults even without a formal anxiety disorder (e.g. shortness of breath, trembling).

The Geriatric Anxiety Inventory (GAI; Pachana et al., 2006) is one of the few anxiety measures designed for older adults. There is preliminary evidence to support its internal consistency, test–retest reliability, and convergent validity with normal older adults and those seeking psychiatric care (Edelstein et al., 2008). The scale employs a dichotomous item format, and the authors minimized the number of somatic items. The Beck Anxiety Inventory (BAI; Beck & Steer, 1990) appears to be a useful and valid measure of anxiety in older adults (Morin et al., 1999; Segal, Coolidge, & Hersen, 1998; Wetherell & Arean, 1997), as does The Penn State Worry Questionnaire (PSWQ; Meyer, Miller, Metzger, & Borkovec, 1990; see also, Beck, Stanley, & Zebb, 1995; Novy et al., 2001).

The Clinical Anxiety Inventory (CAS; Snaith, Baugh, Clayden, Husain, & Sipple, 1982) is a six-item scale derived from the Hamilton Anxiety Scale (HAM; Hamilton, 1959) and has been found to be predictive of both presence and severity of anxiety symptoms in older adults even below diagnostic threshold (Raue et al., 2001). Wisocki, Handen, and Morse's (1986) measure of worry was developed for home-bound and community-dwelling older adults, and should be relevant for some individuals in long-term care. The fear survey developed by Kogan and Edelstein (2004) is content valid for older adults and includes an index of impact on daily activities.

Novy et al. (2001) translated 12 popular well-validated anxiety measures such as the PSWQ (Meyer et al., 1990) and the Fears Questionnaire (FQ; Marks & Mathews, 1979) from English into Spanish and conducted a psychometric evaluation of the translated scales. They also conducted a psychometric evaluation of previously translated scales, such as the BAI – II (Beck & Steer, 1990). They found preliminary support for the validity of use of these translated scales with Spanish-speaking adults (not specifically older adults); however, they noted a significant limitation of the study in the use of highly educated Spanish-speaking adults in the validation sample. See Novy et al. for a complete list of normed measures and applicable psychometric data.

Suicide Assessment. Older adults have the highest rate of suicide in the country (National Institute of Mental Health, 2003). Most older adults who commit suicide are depressed, although the opposite is not the case in that depression does not always precipitate suicidal thinking. Regardless, depressed older adults should be assessed for suicide risk. A variety of variables are associated more highly with suicidal behaviors in older adults, including male (especially White) sex, recent loss of spouse, living alone, chronic or terminal illness or chronic pain, lower education and income, substance use, presence of mental illness, and history of suicide attempt, among others. There are many assessment measures clinicians can use to measure suicidal ideation or behaviors (see Brown, 1999, for a review). However, only three have been developed specifically for older adults.

The Harmful Behaviors Scales (HBS; Draper, Brodaty, Low, & Richards, 2003; Draper et al., 2002) was developed in Australia for use with older adult nursing-home residents. This is a 20-item rating scale that includes the following factors: disorganized behavior, passive self-harm, risk taking, active self-harm, and uncooperativeness. The Reasons for Living scale-Older Adults version

(RFL-OA; Edelstein, McKee, & Martin, 1999) is a 69-item self-report scale that requires the individual to rate various reasons for living (adaptive beliefs), and not taking one's life. The Geriatric Suicide Ideation Scale (GSIS; Heisel & Flett, 2006) is a 31-item measure of suicide ideation, death ideation, loss of personal and social worth, and perceived meaning in life. Both the RFL-OA and GSIS require a moderate level of cognitive functioning. Clinicians should bear in mind that consideration only of demographic and other risk factors, or only the results from assessment instruments is not adequate for gauging a person's risk of suicidal behaviors and do not substitute for a careful evaluation of suicide risk.

Alcohol Abuse Assessment. While some older adults do engage in illegal drug use, such as marijuana, cocaine, and amphetamines, most assessment measures are designed to detect alcohol abuse and dependence. Furthermore, most measures are designed for detection of alcohol abuse problems in the younger adult population. Only the Michigan Alcohol Screening Test – Geriatric version (MAST-G) (Blow et al., 1992) was designed specifically to detect substance abuse problems among older adults. The CAGE (Mayfield, 1974), although not specifically designed for older adults, appears to be suitable for use in this population. The AUDIT (Saunders & Aasland, 1987) and the Cyr and Wartman (1988) may be less useful for use with geriatric patients because of their emphasis on the quantity or frequency of alcohol consumed, which is not a reliable indicator of substance-use problems in the elderly because of the reduced amount of alcohol necessary to achieve deleterious effect (see Beullens & Aertgeerts, 2004). Additional measures designed specifically for older adults have been created (e.g., the UCLS; Luttrell et al., 1997), but further validation is needed.

Assessment of Social Functioning. The assessment of social functioning can be extremely important when considering the mental and physical health of older adults (cf., Burman & Margolin, 1992; Thomas, Goodwin, & Goodwin, 1985). As with younger adults, positive social interactions and good-quality social support can enhance physical and emotional functioning (Oxman & Berkman, 1990). Social support is thought to protect individuals from many of the negative effects of stress. Conversely, negative interactions and poor quality social support can lead to diminished physical and emotional functioning (Rook, 1990). Importantly, Rook suggests that the negative aspects of some human interactions can cancel or out-weigh the benefits of the positive aspects of relationships. The clinician should not always assume that more social interactions are better than fewer.

Relationship patterns and motivations for social interactions change with age. Carstensen (1995) suggests that social interactions can be motivated by information seeking, self-concept, and emotional regulation, with each of these factors weighing in differently at different ages. She has found that older adults are more likely to seek emotional regulation by careful selection of those with whom they interact. Thus, whereas younger adults may prefer a large network of friends and acquaint-ances, the reduced size of an older adult's social network may in some cases contribute to well-being through a concentration of rewarding friendships.

Numerous instruments have been used to assess social relationships and support among older adults. They can be helpful in examining facets of both negative and positive social interactions. These include, for example, the Social Support Structural Interview (Okun, Melichar, & Hill, 1990), the Arizona Social Support Interview Schedule (Barrera, Sandler, & Ramsay, 1981), and the Frequency of Interactions Inventory (Stephens, Kinney, Norris, & Ritchie, 1987). Each of these instruments measures somewhat different aspects of social support. In addition, Northrop and Edelstein (1998) developed a scale measuring assertive behavior in older adults, an important skill in developing and maintaining effective social support and functional relationships. This measure focuses on situations requiring assertive behavior that are likely to be encountered by older adults (e.g., communicating with a physician, turning down a request for a loan). These instruments vary in their reliability, may require considerable subjective judgment, and may be time-consuming for both the interviewer and the participant. Nevertheless, even if standardized assessment is not possible, social variables are important components of the multidimensional assessment, because of their empirically supported relations with physical and emotional functioning.

Environmental Assessment. SNFs, sometimes referred to as "rest homes," are often anything but restful. They are busy, sometimes chaotic, places that may house hundreds of individuals under one roof and employ more scores. Though often less crowded and more home-like, assisted living facilities (ALFs) may still be perceived as communal, institutional environments, very different from the environment older adults are accustomed to. Most individuals in long-term care settings have roommates, often not of their own choice, who can contribute positively or negatively to their well-being. Residents and staff may be subjected to a myriad of noises, including the pings and buzzes of call bells, intrusive overhead paging systems (e.g., "Cleanup in room 313!"), and even the cries and other vocalizations of individuals who are frightened, confused, or in pain. Unwelcome noises can add to agitation and stress, and can interfere with assessment, psychotherapy, and even with simple conversation. Similarly, visual and auditory stimuli abound, especially in the SNF. Lights are often too bright or glaring for older eyes. Even in the most well-maintained facilities, odors of urine, feces, and strong cleaning agents can be offensive. Most modern long-term care settings are designed to maximize physical safety (e.g., appropriate grab bars, hand rails, and nonskid flooring). However, less common are environments that contain the signage and other cues found useful in aiding older adults who may be memory impaired or disoriented.

An analysis of the physical and sensory environment of the older adult is an important element of the multidimensional assessment. The presence or absence of certain stimuli may be related to the symptoms or behaviors that triggered a psychological referral. Even when drastic changes in the physical environment are unfeasible (e.g., call bells may be the only way to alert staff; residents may not be able to afford single rooms), identifying environmental antecedents to problems may be critical in understanding how to manage or eliminate the problem. Administrators and managers may be motivated to improve the environment as a result of a psychological assessment that points to the impact of the environment on behavior and mental health.

Summary

The assessment of older adults in long-term care is often a very complex and challenging endeavor that involves the consideration and integration of multiple elements of information. The complexity of the assessment process is magnified by the need to simultaneously attend to physical- and mental-health problems, the medications used to treat these problems, the interactions and adverse effects of these medications, the normative and nonnormative age-related cognitive and physical decline of older adults, and the interaction of all these factors.

There are also factors that are more likely to require one's assessment expertise in long-term care settings (e.g., capacity to make decisions and consent to treatment, self-destructive behaviors). Moreover, the limited cognitive and functional skills of many residents often requires nontraditional approaches to assessment that test the ingenuity of the clinician. It is the gratification of assembling and integrating all of the relevant clinical variables into a meaningful picture that benefits the residents and aids the multiple disciplinary staff in providing the best possible care that makes the challenges of assessment all very satisfying.

References

Albert, M. (1994). Brief assessments of cognitive function in the elderly. In M. P. Lawton & J. A. Teresi (Eds.), *Annual review of gerontology and geriatrics: Focus on assessment techniques* (pp. 93–106). New York: Springer.

Alexopoulos, G. S., Abrams, R. C., Young, R. C., & Shamoian, C. A. (1988). Cornell scale for depression in dementia. *Biological Psychiatry, 23*, 271–284.

Alexopoulos, G. S., & Mattis, S. (1991). Diagnosing cognitive dysfunction in the elderly: Primary screening tests. *Geriatrics, 46*, 33–44.

American Bar Association Commission on Law and Aging & American Psychological Association. (2005). *Assessment of older adults with diminished capacity: A handbook for lawyers.* Washington, DC: American Bar Association and American Psychological Association.

Areán, P. A., & Miranda, J. (1997). The utility of the Center for Epidemiological Studies-Depression Scale in older primary care patients. *Aging and Mental Health, 1*, 47–56.

Arnold, B. R., Cuellar, I., & Guzman, N. (1998). Statistical and clinical evaluation of the Mattis Dementia Rating Scale – Spanish adaptation: An initial investigation. *Journals of Gerontology: Series B: Psychological Sciences & Social Sciences, 53B*, P364–P369.

Babcock, R. L., & Salthouse, T. A. (1990). Effects of increased processing demands on age differences in working memory. *Psychology and Aging, 5*, 421–428.

Bagby, R. M., Ryder, A. G., Schuller, D. R., & Marshall, M. B. (2004). The Hamilton depression rating scale: Has the gold standard become a lead weight? *American Journal of Psychiatry, 161*, 2163–2177.

Baird, A. D., Ford, M., & Podell, K. (2007). Ethnic differences in functional and neuropsychological test performance in older adults. *Archives of Clinical Neuropsychology, 22*(3), 309–318.

Baker, R. R., Caroll, J. D., Frazer, D. W., Gallagher-Thompson, D., Graca, J., Lichtenberg, P., et al. (1997). *Assessment of competency and capacity of the older adult: A practice guideline for psychologists.* Milwaukee, WI: Department of Veterans.

Baker, F. M., & Espino, D. V. (1997). A Spanish version of the Geriatric Depression Scale in Mexican-American elders. *International Journal of Geriatric Psychiatry, 12*, 21–25.

Barrera, M., Jr., Sandler, I. N., & Ramsay, T. B. (1981). Preliminary development of a scale of social support: Studies on college students. *American Journal of Community Psychology, 9*, 435–447.

Beck, J. G., Stanley, M. A., & Zebb, B. J. (1995). Psychometric properties of the Penn State Worry Questionnaire in older adults. *Journal of Clinical Geropsychology, 1*, 33–42.

Beck, A. T., & Steer, R. A. (1990). Manual for the Beck Anxiety Inventory. San Antonio, TX: Psychological Corporation.

Beck, A. T., Steer, R. A., & Brown, G. K. (1996). *Manual for the revised Beck Depression Inventory.* San Antonio, TX: Psychological Corporation.

Beck, A. T., Ward, C. H., Mendelson, M., Mock, J., & Erbaugh, J. (1961). An inventory for measuring depression. *Archives of General Psychiatry, 4*, 561–571.

Beullens, J., & Aertgeerts, B. (2004). Screening for alcohol abuse and dependence in older people using DSM criteria: A review. *Aging & Mental Heath, 8*, 76–82.

Bland, R. C., & Newman, S. C. (2001). Mild dementia or cognitive impairment: The Modified Mini-Mental State Examination (3MS) as a screen for dementia. *Canadian Journal of Psychiatry, 46*, 506–510.

Blank, K., Gruman, C., & Robison, J. T. (2004). Case-finding for depression in elderly people: Balancing ease of administration with validity in varied treatment settings. *Journals of Gerontology: Biological & Medical Sciences, 59*, 378–384.

Blazer, D. G. (1994). Epidemiology of late life depression. In L. S. Schneider, C. F. Reynolds, III, B. D. Lebowitz, & A. J. Friedhoff (Eds.), *Diagnosis and treatment of depression in late life: Results of the NIH Consensus Development Conference* (pp. 9–19). Washington, DC: American Psychiatric Press.

Blazer, D., George, L. K., Hughes, D. (1991). Anxiety in the elderly: Treatment and research. In C. Salzman, B. D. Lebowitz (Eds.), *The epidemiology of anxiety disorders: An age comparison* (pp. 17–30). New York: Springer.

Blow, F. C., Brower, K. J., Schulenberg, J. E., Demo-Dananberg, L. M., Young, J. P., & Beresford, T. P. (1992). The Michigan Alcoholism Screening Test – Geriatric Version (MAST-G): A new elderly-specific screening instrument. *Alcoholism: Clinical and Experimental Research, 16*, 372.

Borson, S., Scanlan, J., Brush, M., Vitaliano, P., & Dokmak, A. (2000). The Mini-Cog: A cognitive 'vital signs' measure for dementia screening in multi-lingual elderly. *International Journal of Geriatric Psychiatry, 15*, 1021–1027.

Brown, G. K. (1999). *A review of suicide assessment measures for intervention research with adults and older adults.* Retrieved June 14, 2005 from www.nimh.nih.gov/suicideresearch/adultsuicide.pdf

Burman, B., & Margolin, G. (1992). Analysis of the association between marital relationships and health problems: An interactional perspective. *Psychological Bulletin, 112*, 39–63.

Burns, B. J., Wagner, H. R, Taube, J. E., Magaziner, J., Permutt, T., & Landerman, L. R. (1993). Mental health service use by the elderly in nursing homes. *American Journal of Public Health, 83*, 331–337.

Carstensen, L. L. (1995). Evidence for a life-span theory of socio-emotional selectivity. *Current Directions in Psychological Science, 4*, 151–156.

Cassidy, F., & Carroll, B. J. (2002). Vascular risk factors in late onset mania. *Psychological Medicine, 32*, 359–362.

Cheng, S. T., & Chan, A. C. M. (2004). A brief version of the Geriatric Depression Scale for the Chinese. *Psychological Assessment, 16*, 182–186.

Cyr, M. G., & Wartment, S. A. (1988). The effectiveness of routine screening questions in the detection of alcoholism. *Journal of the American Medical Association, 259*, 51–54.

de Craen, A. J. M., Heeren, T. J., & Gussekloo, J. (2003). Accuracy of the 15-item Geriatric Depression Scale (GDS-15) in a community sample of the older old. *International Journal of Geriatric Psychiatry, 18*, 63–66.

Depp, C. A., & Jeste, D. V. (2004). Bipolar disorder in older adults: A critical review. *Bipolar Disorders*, 6(5), 343–367.

Draper, B., Brodaty, H., Low, L.-F., & Richards, V. (2003). Prediction of mortality in nursing home residents: Impact of passive self-harm behaviors. *International Psychogeriatrics*, 15(2), 187–196.

Draper, B., Brodaty, H., Low, L.-F., Richards, V., Paton, H., & Lie, D. (2002). Self-destructive behaviors in nursing home residents. *Journal of the American Geriatrics Society*, 50, 354–358.

Eastham, J. H., Jeste, D. V., & Young, R. C. (1998). Assessment and treatment of bipolar disorder in the elderly. *Drugs & Aging*, 12, 205–224.

Edelstein, B., Goodie, J., & Martin, R. (2004). Physiological and behavioral concomitants of aging. In E. W. Craighead & C. B. Nemeroff (Eds.), *Concise encyclopedia of psychology and neuroscience* (pp. 700–702). New York: Wiley.

Edelstein, B., Martin, R., & Goodie, J. (2000). Considerations for older adults. In M. Hersen & M. Biaggio (Eds.), *Effective brief therapies: A clinician's guide* (pp. 433–448). New York: Academic.

Edelstein, B., Martin, R., & Koven, L. (2003). Assessment in geriatric settings. In J. R. Graham & J. A. Naglieri (Eds.), *Comprehensive handbook of psychology: Volume 10: Assessment psychology* (pp. 389–414). New York: Wiley.

Edelstein, B. A., McKee, D. R., & Martin, R. R. (2000, September). *Development of the Reasons for Living Scale for Older Adults: A suicide assessment instrument.* Poster presented at the Harvard Symposium on Future Research Trends and Opportunities in Aging. Boston, MA.

Edelstein, B., Woodhead, E., Segal, D., Heisel, M., Bower, E., Lowery, A., et al. (2008). Older adult psychological assessment: Current instrument status and related considerations. *Clinical Gerontologist.*

Folstein, M. F., Folstein, S. E., & McHugh, P. R. (1975). Mini-mental state: A practical method for grading the cognitive state of patients for the clinician. *Journal of Psychiatric Research*, 12, 189–198.

Frazer, D. W., Leicht, M. L., & Baker, M. D. (1996). Psychological manifestations of physical disease in the elderly. In L. L. Carstensen, B. A. Edelstein, & L. Dornbrand (Eds.), *The practical handbook of clinical gerontology* (pp. 217–235). Thousand Oaks, CA: Sage.

Gallagher, D., Breckenridge, J., Steinmetz, J., & Thompson, L. (1983). The Beck Depression Inventory and research diagnostic criteria: Congruence in an older population. *Journal of Consulting & Clinical Psychology*, 51, 933–934.

Greenberg, L., Lantz, M. S., Likourezos, A., Burack, O. R., Chichin, E., & Carter, J. (2004). Screening for depression in nursing home palliative care patients. *Journal of Geriatric Psychiatry and Neurology*, 17, 212–218.

Grisso, T. (2002). *Evaluating competencies: Forensic assessments and instruments (Perspectives in law and psychology).* New York: Kluwer.

Hamilton, M. (1959). The assessment of anxiety state by rating. *British Journal of Medical Psychology*, 32, 50–55.

Hamilton, M. (1960). A rating scale for depression. *Journal of Neurology, Neurosurgery & Psychiatry*, 23, 56–61.

Hays, P. A. (1996). Culturally responsive assessment with diverse older clients. *Professional Psychology: Research & Practice*, 27, 188–193.

Heisel, M. J., & Flett, G. L. (2006). The development and initial validation of the Geriatric Suicide Ideation Scale. *The American Journal of Geriatric Psychiatry*, 14(9), 742–751.

Hohl, U., Grundman, M., Salmon, D. P., Thomas, R. G., Thal, L. J. (1999). Mini-Mental State Examination and Mattis Dementia Rating Scale performance differs in Hispanic and non-Hispanic Alzheimer's disease patients. *Journal of the International Neuropsychological Society*, 5, 301–307.

Holtzer, R., Burright, R. G., & Donovick, P. J. (2001). Mattis Dementia Rating Scale: Performance of the very elderly. *Clinical Gerontologist*, 24, 107–114.

Hoyl, M. T., Alessi, C. A., Harker, J. O., Josephson, K. R., Pietruszka, F. M., Koelfgen, et al. (1999). Development and testing of a five-item version of the Geriatric Depression Scale. *Journal of the American Geriatrics Society*, 47, 873–878.

Husain, M. M., Rush, A. J., Sackeim, H. A., Wisniewski, S. R., McClintock, S. M., Craven, N., et al. (2005). Age-related characteristics of depression: A preliminary STAR*D report. *American Journal of Geriatric Psychiatry, 13*, 852–860.

Hyer, L., & Blount, J. (1984). Concurrent and discriminant validities of the Geriatric Depression Scale with older psychiatric inpatients. Psychological Reports, 54, 611–616.

Isella, V., Villa, M. L., & Appollonio, I. M. (2001). Screening and quantification of depression in mild-to-moderate dementia through the GDS short forms. *Clinical Gerontologist, 24*, 115–125.

Janik, S. W., & Wells, K. S. (1982). Multidimensional assessment of the elderly client: A training program for the development of a new specialist. *Journal of Applied Gerontology, 1*, 45–52.

Jefferson, A. L., Powers, D. V., & Pope, M. (2000). Beck Depression Inventory – II (BDI-II) and the Geriatric Depression Scale (GDS) in older women. *Clinical Gerontologist, 22*, 3–12.

Jorm, A. F. (1997). Methods of screening for dementia: A meta-analysis of studies comparing an informant questionnaire with a brief cognitive test. *Alzheimer Disease & Associated Disorders, 11*, 158–162.

Jorm, A. F. (2004). The Informant Questionnaire on Cognitive Decline in the Elderly (IQCODE): A review. *International Psychogeriatrics, 16*, 275–293.

Jorm, A. F., & Jacomb, P. A. (1989). The Informant Questionnaire on Cognitive Decline in the Elderly (IQCODE): Socio-demographic correlates, reliability, validity and some norms. *Psychological Medicine, 19*, 1015–1022.

Jorm, A. F., Scott, R., Cullen, J. S., & MacKimmon, A. J. (1991). Performance of the Informant Questionnaire on Cognitive Decline in the Elderly (IQCODE) as a screening test for dementia. *Psychological Medicine, 21*, 785–790.

Kasl-Godley, J. E., Gatz, M., & Fiske, A. (1998). Depression and depressive symptoms in old age. In I. H. Nordhus, G. R. VandenBos, S. Berg, & P. Fromholt (Eds.), *Clinical Geropsychology* (pp. 211–217). Washington, DC: American Psychological Association.

Kaszniak, A. W. (1990). Psychological assessment of the aging individual. In J. E. Birren & K. W. Schaie (Eds.), *Handbook of the psychology of aging* (3rd ed., pp. 427–445). New York: Academic.

Katz, S., Downs, T. D., Cash, H. R., & Gratz, R. C. (1970). Progress in development of the index of ADL. *The Gerontologist, 10*, 20–30.

Kogan, J., & Edelstein, B. (2004). Modification and psychometric examination of a self-report measure of fear in older adults. *Journal of Anxiety Disorders, 18*, 397–409.

La Rue, A. (1992). *Aging and neuropsychological Assessment*. New York: Plenum.

Lee, D. Y., Lee, K. U., Lee, J. H., Kim, K. W., Jhoo, J. H., Kim, S. Y. (2004). A normative study of the CERAD neuropsychological assessment battery in the Korean elderly. *Journal of the International Neuropsychological Society, 10*, 72–81.

Lesher, E. L. (1986). Validation of the Geriatric Depression Scale among nursing home residents. *Clinical Gerontologist, 4*, 21–28.

Lewinsohn, P. M., Seeley, J. R., Allen, N. B., & Roberts, R. E. (1997). Center for Epidemiologic Studies Depression Scale (CESD-D) as a screening instrument for depression among community-residing older adults. *Psychology and Aging, 12*, 277–287.

Lezak, M. D. (1995). *Neuropsychological assessment* (3rd ed.). New York: Oxford.

Light, L. L. (1992). The organization of memory in old age. In F. I. M. Craik & T. A. Salthouse (Eds.), *Emergent theories of aging* (pp. 111–165). New York: Springer

Lorentz, W. J., Scanlan, J. M, & Borson, S. (2002). Brief screening tests for dementia. *Canadian Journal of Psychiatry, 47*, 723–733

Lowenstein, D. A., Amigo, E., Duara, R., Guterman, A., Hurwitz, D., Berkowitz, N., et al. (1989). A new scale for the assessment of functional status in Alzheimer's disease and related disorders. *Journal of Gerontology, 4*, 114–121.

Lucas, J. A., Ivnik, R. J., Smith, G. E., Ferman, T. J., Willis, F. B., Petersen, R. C., et al. (2005). A brief report on WAIS-R normative data collection in Mayo's Older African Americans Normative Studies. *The Clinical Neuropsychologist, 19*(2), 184–188.

Lukatela, K. C., Ronald, A., Kessler, H., Jenkins, M. A., Moser, D. J., Stone, W. F., et al. (2000). Dementia Rating Scale performance: A comparison of vascular and Alzheimer's dementia. *Journal of Clinical & Experimental Neuropsychology, 22,* 445–454.

Luttrell, S., Watkin, V., Livingston, G., Walker, Z., D'Ath, P., Patel, P., et al. (1997). Screening for alcohol misuse in older people. *International Journal of Geriatric Psychiatry, 12,* 1151–1154.

Mackinnon, A., McCallum, J., Andrews, G., & Anderson, I. (1998). The Center for Epidemiological Studies Depression Scale in older community samples in Indonesia, North Korea, Myanmar, Sri Lanka, and Thailand. *Journals of Gerontology, 53B,* P343–P352.

MacNeill, S., & Lichtenberg, P. (1999). Screening instruments and brief batteries for assessment of dementia. In P. Lichtenberg (Ed.), *Handbook of assessment in clinical gerontology* (pp. 417–441). New York: Wiley.

Marks, I. M., & Mathews, A. M. (1979). Brief standard self-rating for phobic patients. *Behaviour Research and Therapy, 17,* 263–267.

Mattis, S. (1988). *The Dementia Rating Scale: Professional manual.* Odessa, FL: Psychological Assessment Resources.

Mayfield, D. G. (1974). The CAGE-questionnaire: Validation of a new alcoholism-screening instrument. *American Journal of Psychiatry, 131,* 1121–1123.

Meyer, T. J., Miller, M. L., Metzger, R. L., & Borkovec, T. D. (1990). Development and validation of the Penn State Worry Questionnaire. *Behaviour Research and Therapy, 28,* 487–495.

Montgomery, S. A., & Asberg, M. (1979). A new depression scale designed to be sensitive to change. *British Journal of Psychiatry, 134,* 382–389.

Morin, C. M., Landreville, P., Colecchi, C., McDonald, K., Stone, J., & Ling, W. (1999). The Beck Anxiety Inventory: Psychometric properties with older adults. *Journal of Clinical Geropsychology, 5,* 19–29.

Morris, J. C., Heyman, A., Mohs, R. C., Hughes, J. P., Van Bell, G., Fillenbaum, G., et al. (1989). The Consortium to Establish a Registry for Alzheimer's Disease (CERAD). *Neurology, 39,* 1159–1165.

Morrison, J. (1997). *When psychological problems mask medical disorders: A guide for psychotherapists.* New York: Guilford

Moye, J., Gurrera, R., Karel, M., Edelstein, B., & O'Connell, C. (2006). Empirical advances in the assessment of the capacity to consent to medical treatment: Clinical implications and research needs. *Clinical Psychology Review, 26*(8), 1054–1077.

Mui, A. C., Burnette, D., & Chen, L. M. (2001). Cross-cultural assessment of geriatric depression: A review of the CES-D and GDS. *Journal of Mental Health and Aging, 7,* 137–164.

Mui, A. C., & Shibusawa, T. (2003). Japanese American elders and the Geriatric Depression Scale. *linical Gerontologist, 26,* 91–103.

Müller-Thomsen, T., Arlt, S., Mann, U., Maβ, R., & Ganzer, S. (2005). Detecting depression in Alzheimer's disease: Evaluation of four different scales. *Archives of Clinical Neuropsychology, 20,* 271–276.

Nasreddine, Z. S, Phillips, N. A, Bédirian, V., Charbonneau, S., Whitehead, V., Collin, I., et al. (2005). The Montreal Cognitive Assessment (MoCA): A brief screening tool For mild cognitive impairment. *Journal of the American Geriatrics Society, 53,* 695–699.

National Institute of Mental Health. (2003). *Older adults: Depression and suicide facts.* Retrieved June 14, 2005 from http://www.nimh.nih.gov/publicat/elderlydepsuicide.cfm

Norris, J. T., Gallagher, D. E., Wilson, A., & Winograd, C. H. (1987). Assessment of depression in geriatric medical outpatients: The validity of two screening measures. *Journal of the American Geriatrics Society, 35,* 989–995.

Northrop, L., & Edelstein, B. (1998). An assertive behavior competence inventory for older adults. *Journal of Clinical Geropsychology, 4,* 315–332.

Novy, D. M, Stanley, M. A., Averill, P., & Daza, P. (2001). Psychometric comparability of English- and Spanish-language measures of anxiety and related affective symptoms. *Psychological Assessment, 13,* 347–355.

Nuevo, R., Mackintosh, M., Gatz, M., Montorio, I., & Wetherell, J. L. (2007). A test of the measurement invariance of a brief version of the Penn State Worry Questionnaire between American and Spanish older adults. *International Psychogeriatrics, 19*, 89–104.

Okun, M., Melichar, J. F., & Hill, M. D. (1990). Negative daily events, positive and negative social ties, and psychological distress among older adults. *The Gerontologist, 224*, 193–199.

Olin, J. T., Schneider, L. K., Eaton, E. M., Zemansky, M. F., & Pollock, V. E. (1992). The Geriatric Depression Scale and the Beck Depression Inventory as screening instruments in an older adult outpatient population. *Psychological Assessment, 4*, 190–192.

Oxman, T. E., & Berkman, L. F. (1990). Assessments of social relationships in the elderly. *International Journal of Psychiatry in Medicine, 21*, 65–84.

Pachana, N. A., Byrne, G. J., Siddle, H., Koloski, N., Harley, E., & Arnold, E. (2006). Development and validation of the Geriatric Anxiety Inventory. *International Geriatrics* (Online), Jun 29, 1–12.

Patterson, C. J. S., Gauthier, S., Bergman, H., Cohen, C. A., Feighner, J. W., Fleldman, H., et al. (1999). The recognition, assessment and management of dementing disorders: Conclusions from the Canadian consensus conference on dementia. *Canadian Medical Association Journal, 160*, 1S–15S.

Pfeiffer, E. (1975). A short portable mental status questionnaire for the assessment of organic brain deficit in elderly patients. *Journal of the American Geriatrics Society, 23*, 433–441.

Radloff, L. S. (1977). The CES-D Scale: A self-report depression scale for research in the general population. *Applied Psychological Measurement, 1*, 385–401.

Randolph, C., Tierney, M. C., Mohr, E., & Chase, T. N. (1998). The Repeatable Battery for the Assessment of Neuropsychological Status (RBANS): Preliminary clinical validity. *Journal of Clinical & Experimental Neuropsychology, 20*, 310–319.

Rapp, S. R., Parisi, S. A., Walsh, D. A., & Wallace, C. E. (1988). Detecting depression in elderly medical inpatients. *Journal of Consulting and Clinical Psychology, 56*, 509–513.

Raue, P. J., Alexopoulus, G. S., Bruce, M. L., Klimstra, S., Mulsant, B. H., Gallo, J. J., et al. (2001). The systematic assessment of depressed elderly primary care patients. *International Journal of Geriatric Psychiatry, 16*, 560–569.

Reiger, D. A., Narrow, W. E., Rae, D. S., Manderscheid, R. W., Locke, B. Z., & Goodwin, F. K. (1993). The de facto U.S. mental and addictive disorders service system: Epidemiologic catchment area prospective one-year prevalence rates of disorders and services. *Archives of General Psychiatry, 50*(2), 85–94.

Rook, K. S. (1990). Stressful aspects of older adults' social relationships: an overview of current theory and research. In Stephens, M. A. P., Crowther, J. W., Hobfol, S. E. and Tennenbaum, D. L. (Eds.), *Stress and Coping in Later Life Families* (pp. 173–192). Washington, DC: Hemisphere.

Roth, M., Tym, E., Mountjoy, C. Q., Huppert, F. A., Hendrie, H., Verma, S., et al. (1986). CAMDEX: A standardised instrument for the diagnosis of mental disorder in the elderly with special reference to the early detection of dementia. *British Journal of Psychiatry, 149*, 698–709.

Ruchinskas, R. A., & Curyto, K. J. (2003). Cognitive screening in geriatric rehabilitation. *Rehabilitation Psychology, 48*, 14–22.

Sable, J. A., Dunn, L. B., & Zisook, S. (2002). Late-life depression: How to identify its symptoms and provide effective treatment. *Geriatrics, 57*, 18–35.

Salzman, C. (Ed.). (2005). *Clinical geriatric psychopharmacology* (4th ed.). Baltimore: Williams & Wilkins.

Saunders, J. B., & Aasland, O. G. (1987). *Development of a screening instrument.* Geneva, Switzerland: World Health Organization.

Scanlan, J., & Borson, S. (2001). The Mini-Cog: Receiver operating characteristics with expert and naïve raters. *International Journal of Geriatric Psychiatry, 16*, 216–222.

Scogin, F. R. (1998). Anxiety in old age. In I. H. Nordhus, G. R. VandenBos, S. Berg, & P. Fromholt (Eds.), *Clinical Geropsychology* (pp. 205–209). Washington, DC: American Psychological Association.

Segal, D. L., Coolidge, F. L., & Hersen, M. (1998). Psychological testing of older people. In I. H. Nordhus, G. R. VandenBos, S. Berg, & P. Fromholt (Eds.), *Clinical geropsychology* (pp. 231–257). Washington, DC: American Psychological Association.

Sheikh, J. I., & Yesavage, J. A. (1986). Geriatric Depression Scales (GDS): Recent evidence and development of a shorter version. *Clinical Gerontology, 5*, 165–174.

Snaith, R. P., Baugh, S. J., Clayden, A. D., Husain, A., & Sipple, M. A. (1982). The Clinical Anxiety Scale: An instrument derived from the Hamilton Anxiety Scale. *British Journal of Psychiatry, 141*, 518–523.

Spirrison, C. L., & Pierce, P. S. (1992). Psychometric characteristics of the Adult Functional Adaptive Behavior Scale (AFABS). *The Gerontologist, 32*, 234–239.

Stephens, M. A. P., Kinney, J. M., Norris, V. K., & Ritchie, S. W. (1987). Social networks as assets and liabilities in recovery from stroke by geriatric patients. *Psychology and Aging, 2*, 125–129.

Stiles, P. G. (1998). The geriatric depression Scale: A comprehensive review. *Journal of Clinical Geropsychology, 4*, 89–110.

Storandt, M., Botwinick, J., Danziger, W. L., Berg, L., & Hughers, C. (1984). Psychometric differentiation of mild senile dementia of the Alzheimer type. *Archives of Neurology, 41*, 497–499.

Sunderland, T., Alterman, I. S., Yount, D., Hill, J. L., Tariot, P. N., Newhouse, P. A., et al. (1988). A new scale for the assessment of depressed mood in demented patients. *American Journal of Psychiatry, 145*, 955–959.

Tariq, S. H., Tumosa, N., Chibnal, J. T., Perry, M. H., III, & Morley, J. E. (2006). Comparison of the Saint Louis University Mental Status Examination and the Mini-mental State Examination for dementia and mild neurocognitive disorder – a pilot study. *American Journal of Geriatric Psychiatry, 14*, 900–910.

Teng, E. L., & Chui, H. C. (1987). The modified mini-mental state (3MS) examinations. *Journal of Clinical Psychiatry, 48*, 314–318.

Teng, E. L., Hasegawa, K., Homma, A., Imai, Y., Larson, E., Graves, A., et al. (1994). The cognitive abilities screening instrument (CASI): A practical test for cross-cultural epidemiological studies of dementia. *International Psychogeriatrics, 6*, 45–58.

Thomas, P. D., Goodwin, J. M., & Goodwin, J. S. (1985). Effect of social support on stress-related changes in cholesterol level, uric acid, and immune function in an elderly sample. *American Journal of Psychiatry, 121*, 735–737.

Tombaugh, T. N., & McIntyre, N. J. (1992). The Mini-Mental State Examination: A comprehensive review. *Journal of the American Geriatrics Society, 40,* 922–935.

Tschanz, J. T., Welsh-Bohmer, K. A., Plassman, B. L., Norton, M. C., Wyse, B. W., Breitnerr, J. C. S., et al. (2002). An adaptation of the Modified Mini-Mental State Examination: Analysis of demographic influences and normative data. Neuropsychiatry, *Neuropsychology, and Behavioral Neurology, 15*, 28–38.

Watson, L. C., & Pignone, M. P. (2003). Screening accuracy for late-life depression in primary care: A systematic review. *Journal of Family Practice, 52*, 956–964.

Weeks, K. S., Mc Gann, P. E., Michaels, T. K., & Penninx, B. W. J. H. (2003). Comparing various short-form geriatric depression scales leads to the GDS-5/15. *Journal of Nursing Scholarship, 35*, 133–137.

Wetherell, J. L., & Arean, P. A. (1997). Psychiatric evaluation of the Beck Anxiety Inventory with older medical patients. *Psychological Assessment, 9*, 136–144.

Wetherell, J. L., Lenze, E. J., & Stanley, M. A. (2005). Evidence-based treatment of geriatric anxiety disorders. *Psychiatric Clinics of North America, 28*, 871–896.

Whitbourne, S. K. (1999). The aging individual: *Physical and psychological perspectives*. New York: Springer.

Wisocki, P. A., Handen, B., & Morse, C. (1986). The Worry Scale as a measure of anxiety among home bound and community active elderly. *Behavior Therapist, 9*, 91–95.

Wolfe, R., Morrow, J., & Fredrickson, B. L. (1996). Mood disorders in older adults. In L. L. Carstensen, B. A. Edelstein, & L. Dornbrand (Eds.), *The practical handbook of clinical gerontology* (pp. 274–303). Thousand Oaks, CA: Sage.

Yamada, A. M., Valle, R., Barrio, C., & Jeste, D. (2006). Selecting an acculturation measure for use with Latino older adults. *Research on Aging, 28*(5), 519–561.

Yesavage, J. A., Brink, T. L., Rose, T. L., Lum, O., Huang, V., Adey, M., et al. (1982–1983). Development and validation of a geriatric depression screening scale: *A preliminary report. Journal of Psychiatric Research, 17*, 37–49.

Web sites useful to geriatric and geropsychological assessment

- http://www.apa.org/pi/aging/diminished_capacity_part1.pdf– Assessment of cognitive capacity
- http://www.stanford.edu/~yesavage/GDS.html– Geriatric Depression Scale, multiple languages
- http://www.measurementexperts.org/– Psychometric data, descriptions, and articles on assessment measures

Chapter 4
Treatment Plans

Deborah W. Frazer

Introduction

Treatment plans are the central tool for accountability in the psychotherapy process: accountability to the patients, to the insurance companies, to the referring physicians, to the nursing home or assisted-living facilities, and to us, the therapists. Through this tool, we briefly state in writing why we are using psychotherapy, what we hope to accomplish, how we hope to accomplish it, and when we hope to accomplish it.

Background Material

Qualities of Treatment Plans. Constructing a treatment plan is a therapeutic intervention in itself. As a patient and therapist work toward a plan together, they engage in a forthright discussion of the problem(s), what steps will be required to address the issues of concern, and how will they recognize when the treatment has been successful. The therapist must listen to the older adult's unique expressions of distress: his or her "voice" may be influenced by age cohort, educational level, geographic and language origins, race and ethnicity, sexual orientation, religion, and comfort level talking about "personal" issues or with someone of a different background.

Optimally, the therapist must be able to focus the treatment planning discussion on the dominant issues and to explain the appropriate therapeutic strategies, reflecting and incorporating the person's own verbal style. Both patient and therapist then agree on a clear and objective therapy outcome, and identify how they will recognize when it has been achieved. The statement, clarification, and negotiation of these goals are the beginning of psychotherapy.

Codevelopment of a treatment plan engages the older adult, right from the start, as an equal and responsible partner in the therapeutic process. Some older people, who have not previously been included as full partners in their own health care, or who have experienced prior discrimination in health care, might be surprised, confused or skeptical about such a partnership approach. However, sustained therapeutic

E. Rosowsky et al. (eds.), *Geropsychology and Long Term Care*,
DOI: 10.1007/978-0-387-72648-9_4, © Springer Science+Business Media, LLC 2009

partnership and respect will establish the mutual responsibility that will fuel progress on the treatment plan.

Medicare and other mental health insurance payers (including individual managed care plans) require documentation that the treatment plan has been discussed with, and agreed to by the patient or patient representative (e.g., power of attorney, public guardian). Overall, the construction of, and guidance by, a treatment plan is the essential distinction between psychotherapy and "friendly visiting."

- A treatment plan is dynamic. It should be updated whenever there is a major change of the patient's status, goals, objectives, or the therapist's interventions.
- A treatment plan is integrated with the care plans of other providers. In a psychiatric facility, this is accomplished through interdisciplinary care conferences and a single integrated treatment plan. In nursing homes, therapists face a more difficult challenge to align, coordinate, and integrate the psychotherapy plan with the nursing home's assessment (minimum data set – MDS) and care plan. In assisted-living facilities, documentation requirements vary from state to state and among residential providers; therapists must seek information from the facility about how to best integrate treatment plans into the overall "service plan." Psychotherapists working under Medicare are required to consult with the patient's attending or primary care physician. For outpatients or in-home patients, sending the treatment plan to the attending/primary (with the patient's consent) is an efficient way to maintain this communication.
- A treatment plan has specific, individualized, objective, and measurable patient-centered long- and short-term goals. The plan tells what the patient (not the therapist) will accomplish.

Psychologists who have worked in psychiatric hospitals or mental health clinics are familiar with treatment planning, especially if the Joint Commission for the Accreditation of Healthcare Organizations (JCAHO) approves the site. JCAHO insists on a formal treatment planning process, stressing the need for detailed plans with quantifiable objectives. Many managed care companies and other insurance carriers have followed JCAHO's lead and have adopted its high standard for treatment planning.

Although Medicare does not specify what must be included in a treatment plan, therapists are well advised to develop plans compatible with the JCAHO and managed care standards. This ensures the greatest probability of acceptance of the plan by any insurance carrier.

In their local determinations of coverage (LCDs), Medicare carriers often outline elements they expect to see in the beneficiaries' records. For example, if it is the plan to treat a person with dementia, most carriers' LCD requires there be a "reasonable expectation" of improvement or maintenance of functioning that would otherwise decline absent services; that the beneficiary has the capacity to "actively participate" and is cognitively intact enough to have "meaningful" verbal interaction and develop a therapeutic relationship allowing for effective insight-oriented, behavior modifying or supportive therapy. The frequency of services should be both consistent with practice standards and medically necessary for the beneficiary's given condition. Documentation in the treatment plan should provide evidence that the beneficiary is able to meaningfully participate in treatment.

How to Construct a Treatment Plan

Treatment planning is a logical, step-by-step process that follows naturally from the referral and assessment process. To demonstrate the logical progression, include the referral question (the justification for assessment or treatment) and the diagnosis (the conclusion of the assessment process that should guide the treatment process) at the top of the treatment plan. Then proceed to construct the plan itself.

First, choose the most significant and feasible problems to work on. Many problems are raised during the assessment process – some by the patient, others by family members, paid caregivers, or other treating professionals. The initial treatment plan should narrow the focus of treatment to one or, at most, a few problems that are in need of the therapists' attention. Whatever the reason for the referral it is important to address high-risk problems such as suicidal ideation or planning, including documentation of physician notification.

Second, define each problem in behavioral, i.e., observable terms, including statements made by the older adult. For example, instead of stating, "depression," use "depression as evidenced by tearfulness and complaints of sadness." This behavioral definition can seem tedious to therapists who use diagnosis as shorthand for symptom clusters. However, it can be extremely helpful in clarifying communication with the patient, family, physician, the nursing home or assisted-living facility, and insurance payers.

Third, set long-range goals. These are generally framed as positive statements of what the therapist and patient will be striving toward together. Rather than simply "reducing signs of depression," the goal might be to "increase interest in life and feelings of pleasure." The process of specifying these long-range goals gives the therapist and patient time to develop a joint vision of the positive possibilities in the patient's life. Long-range goals are the framework of hope.

Fourth, set specific, measurable objectives for each goal (these are sometimes considered "short-term goals"). These are the small steps that the patient must accomplish to move toward the long-range goal. These are always written as patient objectives or goals, not therapist objectives. To ensure the proper format, start with the stem, "The patient will…." Many therapists find this and the next step the most onerous, as they reduce the artful, almost mystical, process of psychotherapy to very concrete statements of observable behavior. While seeking and experiencing profound moments with the patient, the therapist may be required to write something rather flat, for example: "patient will verbalize conflicts with and ambivalence toward the deceased loved one."

However concrete these behavioral objectives sound, they do help the therapist to focus the treatment. Target dates for the completion of objectives should be included and new objectives added as previous objectives are met. If an objective is not met after the expected time interval, it should serve as a discussion point between the therapist and patient. Perhaps the objectives should be changed, or the therapist's intervention should be modified.

Fifth, and finally, specify the therapist's interventions and the frequency of therapy. This states what the therapist will do, and how often, that are intended to result in the patient's accomplishing particular objectives. Try the stem, "the therapist will...." to ensure proper format. These interventions can be written from any theoretical orientation. It may require several therapist interventions for the patient to accomplish an objective. If one intervention is not working, substitute another. Be as specific and as jargon-free as possible, so that reviewers can understand the therapeutic interventions.

If the intervention significantly deviates from standard practice – either in type or frequency of intervention – justify the change in approach on the treatment plan. For example, the therapist and care team may decide that short daily sessions are needed for a 2-week period to prevent hospitalization of a patient. Specify clearly on the plan that daily sessions are a short-term therapeutic strategy with a specific goal. It should be added that third-party payers may also look closely at the interventions portion of the treatment plan. When the psychologist's interventions are explicitly stated in the plan and in the ongoing progress notes, there is greater likelihood of coverage for the service by the payer, if the documentation is reviewed or audited. Conversely, when the interventions are vague or does not differentiate between what a licensed mental health practitioner provides and what any other staff person may provide, there is increased risk of denial of the claims. This is an important consideration in the development of comprehensive treatment plans for the older adult residing in a long-term care setting.

Summary

The completed treatment plan, then, consists of six elements specified in writing: diagnosis, identified problem(s), behavioral definitions of problems(s), patient long-range goals, patient objectives or short-term goals, and therapist interventions. Such a plan, developed together with the resident, should serve both the older adult and the therapist well.

References

Bennett, B. E., Bricklin, P. M., Harris, E., Knapp, S., VandeCreek, L., & Younggren, J. N. (2006). *Assessing and managing risk in psychological practice: An individualized approach*. Rockville, MD: The Trust.

Carstensen, L. L., Edelstein, B. A., & Dornbrand, L. (1996). *The practical handbook of clinical gerontology*. Thousand Oaks, CA: Sage.

Carstensen, L. L., & Fisher, J. E. (1991). Treatment applications for psychological and behavioral problems of the elderly in nursing homes. In P. A.Wisocki (Eds.), *Handbook of clinical behavior therapy with the elderly client*. New York: Plenum.

Empire Medicare CMS Local Coverage Determination, LCD for Mental Health Services http://www.empiremedicare.com/newypolicy/policy/L3671_final.htm#31

Frazer, D. W., & Jongsma, A. E. (1999). *The older adult psychotherapy treatment planner*. New York: Wiley.

Hersen, M., & Van Hasselt, V. B. (1992). Behavioral assessment and treatment of anxiety in the elderly. *Clinical Psychology Review, 12*, 619–640.

Joint Commission on Accreditation of Healthcare Organizations. (1999). *1999–2000 Standards for Behavioral Health Care*. Oakbrook Terrace, IL: Joint Commission on Accreditation of Healthcare Organizations.

Lichtenberg, P. A. (1994). *A guide to psychological practice in geriatric long term care*. New York: Haworth.

Niederehe, G., & Schneider, L. S. (1998).Treatments for depression and anxiety in the aged. In P. E.Nathan & J. M.Gorman (Eds.), *A guide to treatments that work*. New York: Oxford.

Teri, L., & Gallagher, D. G. (1991). Cognitive-behavioral interventions for treatment of depression in Alzheimer's patients. *The Gerontologist, 21*, 413–416.

Teri, L., & Logsdon, R. (1991). Identifying pleasant activities for individuals with Alzheimer's disease: The Pleasant Events Schedule-AD. *The Gerontologist, 31*, 124–127.

Chapter 5
Treatment Process

Michael Duffy

Introduction

This chapter provides an overview of several important parameters of providing quality psychological treatment to residents in long-term care settings. Clinical geropsychologists made early important strides in developing psychological assessment procedures for residents in long-term care setting (see Chap. 3) and also now have focused increasing attention on providing treatment services. Newly available resource materials will be cited throughout this chapter.

Work in nursing homes and assisted-living communities is distinctly different from general outpatient office psychotherapy and requires a greater flexibility and tolerance for ambiguity in the therapist. As in hospital and residential settings, older adults do not view themselves as mental health patients, and generally belong to a cohort that has little cultural appreciation for the concept of mental health. Persons who weathered the Great Depression and the Second World War often developed a survival mentality that has little time for the niceties of emotional well-being; happiness is often viewed as a luxury item. We know, however, that many older adults critically need psychological help (Karlin & Duffy, 2004; Karlin, Duffy, & Gleaves, in press) and are usually referred by pastors, family members, and physicians as well as by employees in long-term care facilities. The advent of Medicare reimbursement of psychological services has notably increased treatment of nursing home residents although policy limitations still hamper effective delivery (Karlin & Humphries, 2007).

Background Material

Optimal Treatment Methods. Make sure your diagnostic understanding of the resident is systemic (holistic), that is, it takes into account the various forces and influences that play on this particular client or situation. These will include cultural and ethnic factors, family relations, socioeconomic factors, temperament and personality style, medical conditions, etc. (Duffy, 2002). Often the effective treatment of

E. Rosowsky et al. (eds.), *Geropsychology and Long Term Care*,
DOI: 10.1007/978-0-387-72648-9_5, © Springer Science+Business Media, LLC 2009

depression in the nursing home resident, for example, will involve intervention with adult children (Duffy, 1987).

It is important to take into account DSM-IV-TR Axis II (personality profile) as well as Axis I (symptomatic profile), even if neither reaches the level of a diagnosable disorder (Rosowsky, E., Abrams, R., & Zweig, R. (Eds.), 1999) These domains are rarely separable; treating symptoms without appreciating personality dynamics is often frustrating. However, minimizing depression symptoms in a personality disorder can be life threatening. Comorbidity of chronic medical and psychological conditions is especially common among geriatric patients. Be aware that comorbidity implies not only coexistence of conditions, but also their reciprocal interaction (Sadavoy, 1999). It is likely, for example, that some treatment failures with depression or treatment-resistant depressions are due to unappreciated personality dynamics, which are recip-rocal with the depression. While depressions may be symptomatically similar, there are many internal psychological variations that affect appropriate treatment strategy such as complicated bereavement and sociopathic dejection (Duffy, 2005).

Since diagnoses are complex, treatments will need to be designed for multiple impacts. With an individual client we may need to use behavioral methods for aggres-sive treatment of acute symptoms (Hussian & Davis, 1985; Logsden, McCurry, & Teri, 2007); cognitive, physical, or interpersonal approaches for depression (Coon, Rider, Gallagher-Thompson, & Thompson, 1999); dynamic methods for personality disorders (Jacobowitz & Newton, 1999); and multigenerational therapy for family relational problems (Duffy, 1986). Especially note the recent clinical and research evidence for the effectiveness of physical exercise on depression (Duffy & Karlin, 2006).

Treatment Approaches with Demonstrated Effectiveness. Providing effective psychotherapy services for older adults is clearly of great importance. The APA Division 12 Task Force devoted to establishing an evidentiary base for psycho-therapy, using a set of quality-assured research criteria, developed an inventory of "well established" and "probably efficacious" treatment methods (Chambliss et al., 1996). More recently, Gatz et al. (2001) performed a similar review of treatments designed for older adults, which often require modification for special late-life fac-tors (Satre, Knight, & David, 2006). The Gatz et al. article includes behavioral, cognitive, interpersonal, and short-term psychodynamic approaches to a series of problem scenarios such as depression, sleep disorders, alcoholism, and affective symptoms in dementia patients.

Meta-analyses (Scogin & McElreath, 1994; Pinquart & Sorenson, 2001) have also shown that psychological treatments with older adults are effective. Recent literature reviews (Logsden et al., 2007; Ayers, Sorrell, Thorpe, & Wetherall, 2007) have also added to our knowledge of the range of effective treatments for disruptive behaviors, late-life anxiety, and caregiver stress (Gallagher-Thompson & Coon, 2007).

It transpires that psychological and behavioral treatment with older adults has been found to have similar or superior efficacy than psychotropic medication in reducing depression in older patients (Antonuccio, Danton, & DeNelsky, 1995; Scazufca & Matsuda, 2002). So, bearing in mind the diagnostic complexity and comorbidity mentioned earlier, these geriatric psychotherapy approaches provide the clinician with an important armamentarium in treating older adults.

There are two limitations in this valuable work: (1) in some cases the supporting research was conducted in laboratory or quasi-laboratory settings rather than in clinical settings, and thus the term *efficacious* is used vs. the term "effective" in research conducted in actual clinical settings; (2) to date much of the research is conducted with symptomatic (Axis I) problems with less attention so far to personality problems (Axis II), which are frequently the most challenging problems in the nursing home (Rosowsky & Smyer, 1999).

Problem diagnoses that are complex, personality-related or comorbid, as is frequently the case may not respond as well to treatments supported by research on single diagnoses. An alternate source of information about effective treatments in more complex conditions is the consensus of clinical experience and viewpoints found in the clinical literature. Although lacking experimental design features and large N (rarely possible in clinical settings), these clinical commentaries are rich in intensive case analysis. In the parlance of current nontraditional or qualitative methodologies (which are often compatible with clinical practice), such case descriptions provide "thick" data. These case descriptions become even more pertinent as clinical consensus begins to build in complex diagnostic and treatment issues. There are increasing numbers of resources that present such material (APA, 2001; Duffy, 1999; Knight, 1986).

In addition to discrete psychological techniques and interventions, effective psychotherapy relies on the power of the therapeutic relationship (Norcross, 2001, 2002). Recent work on "common factors" in psychotherapy (Wampold, 2001) suggests that a significant portion of the effectiveness of specific interventions may be accounted for by the nature and quality of the underlying therapeutic relationship with the client. This might include, for example, not only caring and acceptance, but also the less-discussed variable of the "authority" of the therapist.

Therapist authority (a sense of therapist self-efficacy) may be critical, in the effective use of such techniques as thought stopping or relaxation training. Therapist skill in appreciating and strategically using the often-subtle dynamics in the therapeutic relationship with the older client will enhance the effectiveness of behavioral and cognitive techniques. Classic works emphasizing the therapeutic relationship by Harry Stack Sullivan (1968), Carl Rogers (1951) and Bill Kell (1966) are helpful in revisiting the interpersonal process domain in psychotherapy. And as cited above, recent research has reaffirmed that a therapeutic relationship that is "intimate" is a powerful change agent in itself. This occurs where "the person of the therapist" provides a climate in which both resident and therapist can be completely themselves and thus explore critical developmental and interpersonal dimensions of the problem. Oftentimes, anxiety and depression are inevitable signals of internal processes of pain, despair, loss, or a sense of abandonment, which must also become the focus of an intensive interpersonal therapeutic process.

Psychotherapy with residents with dementia will also benefit from both technical and relationship aspects of therapy. Where cognitive impairment was once thought to inhibit psychotherapy and make it redundant, recent understanding of retained emotional life and even emotional memory makes psychotherapy powerful even with residents with dementia (Duffy, 1999, 2005). Even in the face of dementia, there is evidence of continued emotional life, including a continuity of personality and attachment style (Magai, Cohen, Culver, Gomberg, & Malatesta, 1998). It is critical that psychologists help family members and long-term care employees to appreciate the need for continued interaction with dementing residents, who will otherwise show symptoms related of emotional abandonment and comorbid depression (Duffy, 2006). The best therapeutic guideline to all caregivers of older persons with cognitive impairment is "keep talking to them." As with neonatal children, conversation is the conduit for many rich emotional resources in addition to logical meaning and understanding (Duffy, 1999).

The Context of Psychotherapy in Long-Term Care

Psychotherapy in long-term care settings, such as in nursing homes and assisted-living facilities, is quite unlike the peaceful environment of a private practice office. It is common to have no individual conference room space available. Privacy is sometimes found in a borrowed office, a hairdressing salon, or the chapel. Frequently, especially with cognitively impaired residents, we may need to find and interact with them in public spaces, such as an entry lounge, dining room, or we might not be able to provide the service at all. Even when a private space has been located, it is not uncommon to have an assessment or therapy session interrupted by the client's friend, roommate, or curious neighbor. But, as is often the case in life, serendipity can save the day. In many of these difficult cases, we find invaluable collateral information about our client through these "intrusive" occurrences as we watch our client interact with the intruder. What started out as individual therapy becomes group therapy!

Because of the situation described earlier, the traditional understanding of informed consent and confidentiality is frequently challenged (Duffy, 2002; Norris, 2002; Zehr, 2002). Patients will occasionally publicly introduce their "personal doctor" as a trophy, broadcast details of the therapy process, and oftentimes require urgent help before informed consent can be obtained. With regard to confidentiality, it may also be unethical *not* to discuss the case with nurse aides who may (unwittingly) be undoing, between visits, the therapeutic gains and strategies of the psychologist. Both Medicare regulations and the recent HIPAA (Health Insurance Portability and Accountability Act) rules emphasize the importance of consultation and collaboration with associated facility staff. Fortunately the most recent revision of the APA Ethical Principles and Code of Conduct (APA, 2002) recognized these complicated situations often found long-term care settings, as well as acute care hospitals and pediatric settings and present language that is sensitive and flexible. Informed consent, for example, must be gained as soon "as is feasible."

Coordinated Multidisciplinary (Interdisciplinary) Treatment. Because of the reciprocal body–mind relationship, psychological practice should always involve interdisciplinary cooperation. The growing role of psychology in primary care helps emphasize this point. This cooperation is particularly crucial in geriatric health care where many chronic medical conditions coexist with or mimic mental health disorders and vice versa.

Cooperation is especially important in dementia care. Neurological and general medical evaluation is an integral and necessary part of a thorough psychological or neuropsychological assessment.

On a day-to-day basis, psychotherapists can benefit greatly by respectful and flexible relationships with all of the health care professionals (Sanders, Brockway, Ellis, Cotton, & Bredin, 1999) found in long-term care settings. Family physicians, while rarely frequent visitors, are usually open to psychological input, even to general suggestions on the need for psychoactive medication. Psychologists can in turn be much better informed about medical conditions that impact the total picture. Establishing liaisons with nursing assistants is especially important in gaining critical contextual information and can also help gain their adherence to the psychologist's overall therapeutic strategy underway with the resident. Lack of collaboration might well mean the unwitting undermining of the strategy.

Caseload Size and Reimbursement Procedures. The advent of Medicare reimbursement has in some cases encouraged a "factory-like, assembly line" provision of service where sometimes minimally trained psychologists are recruited to provide services with an exclusive focus on "billable hours" (or often, half hours). This raises ethical issues in the practice of psychology and has also brought intensive scrutiny from the Center for Medicare and Medicaid Services (CMS). And, while subsequent CMS restraints are often an overcorrection (since many psychological services are needed in long-term care settings), this alerts geropsychologists to maintain the highest quality of psychological care even in this challenging environment.

Psychotherapists adjust their caseloads to their capacity to provide high-quality and effective therapeutic contact with their resident clients. The goal is to develop a significant and even intimate, life-changing relationship with the client so that maximum therapeutic effectiveness can occur. The length of sessions (and related CPT codes) should reflect the needs and tolerance of the residents. Psychotherapists are also wise to be sensitive to their own energy and tolerance levels and to adjust session length to their own needs as well as the resident's needs. Sitting in uncomfortable position during counseling can distract one's concentration. Some psychologists carry their own light-weight, adjustable chair that can offer reasonable comfort in a variety of settings.

It is critical that therapists include conferring with facility staff in their services – even if not often a reimbursed effort (though procedure codes do exist for team consultations). Too frequently staff remark that Dr. X comes, sees patients, and leaves without seeking any input or giving feedback to staff. Concern for the client will require coordination of treatment strategies, especially with those staff members who make treatment decisions (directors of nursing) and those who care for the patient daily (nurse aides). Current CMS initiatives are likely to allow some reimbursable consultation with staff.

It is critical that psychologists support the efforts of professional psychology organizations in legislative and regulatory reform. It is important that clinical geropsychologists actively seek the expansion of insurance coverage, with appropriate estimates of the relative work values, to include legitimate services such as consultation and case and behavior management, now widely covered in the new Health and Behavior codes developed by CMS. These services are integral to comprehensive care, and adequate reimbursement will help insure their delivery.

Parameters for Discontinuing Treatment. In some cases, it becomes acceptable and even necessary to discontinue treatment with an older client and possibly refer to another professional. This can come about for a variety of reasons. The most obvious, of course, is that the patient has improved sufficiently to discontinue treatment. Other reasons for discontinuing treatment include the following: the progress of the case may require interventions that are beyond our current skill level or specialty, or cognitive decline leaves the resident unable to participate in and benefit from psychological treatment, or there may be a personality incompatibility that inhibits progress. There may be a relocation of either client or therapist. In all these cases adequate closure needs to be reached through appropriate termination or referral.

It is not appropriate to discontinue treatment out of simple personal convenience, dislike of the setting, or dislike of the client. In the case of dislike of the client, especially, clinical supervision or consultation from a colleague should be sought, since these emotions are likely to represent an impasse in the therapeutic process (and possibly signal an opportunity for the "real beginning of therapy") rather than an incidental occurrence. This closer look will also reveal how the client may trigger similar reactions in other caregivers, and thus, point the way toward an additional focus and goal of treatment.

Referrals to other professionals should only occur when the above considerations have been reviewed. Referrals should be followed through with the client and the new caregiver to insure the client's needs are addressed. In all cases, there must be careful avoidance of emotionally abandoning a client. When the therapist's covert intention is not the well-being of the client, even when these ethical steps are taken, the client will often get the subverbal message that they are being "disposed of," leading to feelings of emotional abandonment with all of its consequences. These principles are particularly important in care of the infirm or frail and/or the older adult with progressive cognitive disease. Such cases may provide little gratification to the professional, and so it may be tempting to find a reason to discontinue services. Thoughtful examination of one's motives while maintaining primary focus on the needs of the client can help the therapist ethically navigate through these difficult waters.

Sensitivity in the Process of Ending Treatment. Ending a course of treatment can be difficult in residential care settings; we leave our clients rather than they leave us. This becomes relevant when we have helped alleviate or control the symptomatic problems, or when the personality adjustment problems are at a point of diminishing returns. We know that therapy with people suffering narcissistic or borderline problems, for example, rarely continues uninterrupted through the full therapeutic trajectory. Rather, gains are made through a series of short-term therapeutic periods, with breaks for relief from, for example, boredom with or anger at the therapist.

Ending clinical services is probably best conceived as "closure" and for some a "time out" rather than the less helpful concept and imagery of "termination;" in some sense, relationships that have become significant continue internally within the self as a rich source of life and nourishment. This is particularly helpful in the long-term care context where we look for a sense of "goodbye for now" with our resident clients. Often, we will be back again and even a later friendly visit or corridor recognition and warm greeting can cement the sense of therapeutic gain that has been achieved. We may even, for example, pay a holiday visit to a former client in a natural and intimate way that signals the continued existence of the relationship.

Sometimes psychotherapists are overly vigilant in these circumstances about creating dependency, and paradoxically, by strenuously avoiding it, actually intensify dependency. At such moments therapists might be helped to remember the developmental truth that being allowed to healthily depend nourishes independence and strengthens autonomy. Even in our very old clients, if we feel secure enough to allow our clients to "cling," we will frequently find they no longer need to.

References

American Psychological Association. (2001). *Psychological services in long-term care: Resource guide*. APA Committee on Aging web site (APA.org).

American Psychological Association. (2002). *APA ethical principles of psychologists and code of conduct*. Washington, DC: American Psychological Association.

Antonuccio, D. O., Danton, W. G., & DeNelsky, G. Y. (1995). Psychotherapy versus medication for depression: Challenging the conventional wisdom with data. *Professional Psychology: Research and Practice, 26*(6), 574–585.

Ayers, C. R., Sorrell, J. T., Thorp, S. R., & Wetherall, J. L. (2007). Evidence-based psychological treatments for late-life anxiety. *Psychology and Aging, 22*(1), 8–17.

Chambliss, D. L., Sanderson, W. C., Shoham, V., Johnson, S. B., Pope, K. S., Crits Christoph, P., et al. (1996). An update on empirically validated therapies. *The Clinical Psychologist, 49*(2), 5–18.

Coon, D. W., Rider, K., GallagherThompson, D., & Thompson, L. (1999). Cognitive-behavioral therapy for the treatment of late-life distress. In M. Duffy (Ed.), *Handbook of counseling and psychotherapy with older adults*. New York: Wiley.

Duffy, M. (1986). The techniques and contexts of multigenerational therapy. *Clinical Gerontologist, 5*(3/4), 347–362.

Duffy, M. (1987). The techniques and contexts of multigenerational therapy. *Clinical Gerontologist, 5*(3/4), 347–362.

Duffy, M. (1999). Reaching the person behind the dementia: Treating co-morbid affective disorders through sub vocal and nonverbal strategies. In M. Duffy (Ed.), *Handbook of counseling and psychotherapy with older adults*. New York: Wiley.

Duffy, M. (2002). Confidentiality and informed consent versus collaboration: Challenges in psychotherapy ethics in nursing homes. *The Clinical Gerontologist, 25*(3/4), 277–292.

Duffy, M. (2005). Psychotherapeutic interventions for older persons with dementing disorders. In C. Brody & V. Semel (Eds.), *Strategies for therapy with the elderly: Living with hope and meaning* (2nd Ed.). New York: Springer.

Duffy, M. (2006). Psychotherapeutic interventions for older persons with dementing disorders. In C. M. Brody & V. G. Semel (Eds.), *Strategies for therapy with the elderly: Living with hope and meaning* (2nd ed., pp. 15–37). New York: Springer.

Duffy, M., & Karlin, B. (2006). Treating depression in nursing homes: Beyond the medical model. In L. Hyer & R. C. Intrieri (Eds.), *Long-term care: Psychological and psychosocial assessment and treatment*. New York: Springer.

Gallagher-Thompson, D., & Coon, D. W. (2007). Evidence-based psychological treatments for distress in family caregivers of older adults. *Psychology and Aging, 22*(1), 37–51.

Gatz, M., Fiske, A., Fox, L. S., Kaskie, B., Kasl-Godley, J. E., McCallum, T. J., et al. (2001). *Journal of Mental Health Counseling*

Hussian, R. A., & Davis, R. L. (1985). *Responsive care: Behavioral interventions with elderly persons.* Champaign, IL: Research Press.

Jacobowitz, J., & Newton, N. (1999). Dynamics and treatment of narcissism in later life. In M. Duffy (Ed.), *Handbook of counseling and psychotherapy with older adults* (pp. 453–469). New York: Wiley.

Karlin, B. E., & Duffy, M. (2004). Geriatric mental health policy: Impact on service delivery and directions for effecting change. *Professional Psychology: Research and Practice, 35,* 509–519.

Karlin, B. E., Duffy, M., & Gleaves, D. H. (in press). Patterns and predictors of mental health service use and mental illness among older and younger adults in the United States. *Psychological Services.*

Karlin, B. E., & Humphries, K. (2007). Improving medicare coverage of psychological services for older Americans. *American Psychologist, 62*(7), 637–649.

Kell, B., & Mueller, W. (1966). *Impact and change: A study of counseling relationships.* Englewood Cliffs, NJ: Prentice Hall.

Knight, B. (1986). *Psychotherapy with older adults.* Beverly Hills, CA: Sage.

Logsden, R. G., McCurry, S. M., & Teri, L. (2007). Evidence-based psychological treatments in individuals with dementia. *Psychology and Aging, 22*(1), 28–36.

Magai, C., Cohen, C. I., Culver, C., Gomberg, D., & Malatesta, C. (1998). Relation between premorbid personality and patterns of emotion expression in mid to late-stage dementia. *International Journal of Geriatric Psychiatry, 12*(11), 1092–1099.

Norcross, J.C. (Ed.)(2001) Empirically-supported therapy relationships: Summary report of the Division 29 Task Force. *Psychotherapy, 38, 4.*

Norcross, J. C. (2002). *Psychotherapy relationships that work.* New York: Oxford University Press.

Norris, M. P. (2002). Psychologist's multiple roles in long-term care: Untangling confidentiality quandaries. *The Clinical Gerontologist, 25*(3/4), 261–275.

Pinquart, M., & Sorenson, S. (2001). How effective are psychotherapeutic and other psychosocial interventions with older adults: A metaanalysis. *Journal of Mental Health and Aging, 7,* 207–243.

Rogers, C. (1951). *Client centered therapy.* New York: Free Press.

Rosowsky, E., Abrams, R., & Zweig, R. (Eds.). (1999). Personality disorders in older adults: Emerging issues in diagnosis and treatment. Mahwah, NJ: Erlbaum.

Rosowsky, E., & Smyer, M. (1999). Personality disorders and the difficult nursing home patient. In E. Rosowsky, R. Abrams, & R. Zweig (Eds), *Personality disorders in older adults: Emerging issues in diagnosis and treatment* (pp. 257–274). Mahwah, NJ: Earlbaum.

Sadavoy, J. (1999). The effect of personality disorders on Axis I disorders in the elderly. In M. Duffy (Ed.), *Handbook of counseling and psychotherapy with older adults* (pp. 397–413). New York: Wiley.

Sanders, K., Brockway, J. A., Ellis, B., Cotton, E. M., & Bredin, J. (1999). Enhancing mental health climate in hospitals and nursing homes: *Collaboration strategies for medical and mental health staff.* In M. Duffy (Ed.), Handbook of counseling and psychotherapy with older adults. New York: Wiley.

Satre, D. D., Knight, B. G., & David, S. (2006). Cognitive-behavioral interventions with older adults: Integrating clinical and gerontological research. *Professional Psychology: Research and Practice, 37*(5), 489–498.

Scazufca, M., & Matsuda, C. (2002). Review of the efficacy of psychotherapy vs. pharmacotherapy for depression treatment in old age. *Revista Brasileira de Psiquiatria, 24,* 64–69.

Scogin, F., & McElreath, L. (1994). The efficacy of psychosocial treatments for geriatric depression: A quantitative review. *Journal of Consulting and Clinical Psychology, 62,* 69–74.

Sullivan, H. S. (1968). *The interpersonal theory of psychiatry.* New York: Norton.
Wampold, B. E. (2001). *The great psychotherapy debate: Models, methods, and findings.* Mahwah, NJ: Erlbaum.
Zehr, M. D. (2002). Informed consent in the long term care setting. *The Clinical Gerontologist, 25*(3/4), 239–260.

Chapter 6
Integration of Psychology, Psychiatry, and Medication in Long-Term Care

Lee Hyer and Shalija Shah

Use of long-term care (LTC) facilities has become a necessary alternative in the health armamentarium in all developed countries. In the United States, 24% die in nursing homes (Porock et al., 2005). Although only 5% of individuals over age 65 are living in LTC facilities at any given point in time, nearly 40% will spend at least some portion of their lives there (U.S. Census Bureau, 2002). Annually almost 2 million adults are admitted to one of 16,800 nursing homes in the United States (Hyer & Ragan, 2002; Rhoades & Krauss, 1999). The recent growth of assisted living facilities too – now estimated at over 36,000 facilities serving about one million residents (Stefanacci, 2005) – complements these nursing homes within the spectrum of LTC settings.

Nursing homes themselves are now mini-medical facilities, functioning not as "homes" but as hospitals from the perspective of the residents (Bergman-Evans, 2004). In this regard, LTC environments present residents with many challenges, including lack of privacy, confinement to an institutional schedule, and the knowledge that the LTC setting is likely to be an individual's final home. In this environment, the challenge of integrating mental health into the care of the aging within these varied LTC settings is enormously important. In fact, mental and medical pathologies often share a common expression and function. Physical changes, like those seen in Parkinson's disease, have definite medical etiologies, but frequently fall into the domain of mental health, given their neuropsychological properties and high incidence of comorbid depression (Dening & Bains, 2004).

A majority of residents living in LTC facilities carry a least one psychiatric diagnosis. As many as 80% of nursing home residents will have dementia or another diagnosable psychiatric disorder (Hyer & Ragan, 2002; Rovner et al., 1990). Furthermore, between 6 and 24% have a major depressive disorder diagnosis, 30% have minor depression or dysthymia, and as many as 35% manifest depressive symptoms. Minimum Data Set (MDS) reports that 14.6% of LTC residents have been diagnosed with anxiety disorder, though this rate may be deceptively low given limitations of assessment and in reporting anxiety symptoms within the MDS (Centers for Medicare and Medicaid Studies, 2004).

Despite these high rates of psychiatric disorders, most residents are not under the care of a mental health clinician. Typically, these services are provided by independent psychiatric consultants who see specific residents on an as-needed, on-call basis

E. Rosowsky et al. (eds.), *Geropsychology and Long Term Care*,
DOI: 10.1007/978-0-387-72648-9_6, © Springer Science+Business Media, LLC 2009

(Bartels, Moak, & Dums, 2002). In many cases medication alone is given. The Centers for Medicare and Medicaid Services (Boyle et al., 2004) report a 97% increase in antidepressant medications for all residents from 12.6% to 24.9%. Datto et al. (2002) suggest that 35% of nursing home residents receive antidepressant medication.

The Practice of Mental Health in LTC Matters. Ormel, Van cen Brink, and Koeter (1995) indicated that evidence tells us that there are consequences to a lack of mental health input in these facilities, including (1) higher health care utilization and costs; (2) greater functional impairment; (3) increased utilization of staff time; (4) nonadherence to medical care; (5) increased mortality; and (6) reduced quality of life.

Clinical common sense tells us that both dementia and depression result in increased time for staff, even after controlling for physical illness and disability (Fries et al., 1993). Behavioral health care also makes a difference with medical procedures, as in hip-fracture, where residents receiving psychiatric services experienced fewer complications and were 9 times more likely to resume functioning at preoperative levels (Strain, Lyons, & Hammer, 1991).

In this chapter we will address these issues. We discuss the medical model, as well as psychosocial models. We also provide a typology of residents in LTC for which the practice of psychology is applied. From here, we will address the psychiatric care of residents as viewed from the literature and clinical care in LTC. We then consider the practice of psychology in these settings, both assessment and treatment. Throughout, we argue that psychology, as a profession, makes a difference in LTC environments, and that integrated care is important, if not key, in the overall scheme of healthcare and life quality.

Models of Care

The Most Applied Model in LTC Settings is Medical. The core idea is that there is a pathogen and this can be corrected or alleviated by medical interventions. Becker and Kaufman (1988) discuss why the medical model is appealing in LTC, where aging is considered a physiological event first and foremost. People arrive in a LTC setting for one reason, illness. "Medical" needs are acted upon vigorously. Medical interventions are provided by nurses and custodial care is provided by nurse assistants or aides. This medical care is supplemented by psychiatric input, with medication usage rising measurably in the past 5years.

In spite of this, there is some consensus that nonpharmacological interventions are an appropriate first-line intervention when there are mental or behavioral health problems. After a thorough review, Jacobo, Mintzer, and colleagues (1997) noted that nonpharmacolgic interventions constitute the basis for dementia care. The issue is how to have a care plan that is integrated, clear, accepted by all members of the treatment team, monitored, evaluated, and consistently applied. Psychopharamacology alone is problematic and is the source of many avoidable untoward events and costs.

Perhaps the most appealing model in LTC is biopsychosocial, chiefly the person–environment interaction model in LTC settings. Kahana and Kahana (1982) warned that we must attend to individual differences or we may unintentionally increase the incongruence between person and environment, thus creating further stress. Carp and Carp (1984) also lamented that the relationship between individual and context is critical and, if out of alignment, encourages debility.

The seminal understanding of the resident in LTC has come, however, from Powell Lawton (1999). He held that the residential settings of older adults must fulfill three standard functions; maintaining competence, providing stimulation, and providing a sense of security and support. The goal of LTC settings is to keep residents as independent and autonomous as possible, in a comfortable environment. The person–environment fit model is designed to stabilize and build on the competencies of the residents, being both creative and supportive, yet challenging so that the potential for prolonged functioning exists.

More recently, there are several models applied in LTC that offer explanations for challenging behaviors (Cohen-Mansfield, 2001). First, the "unmet needs" model proposes that challenging behaviors occur when there is a deficit in the environment leading the resident to become bored, lonely, or deprived of sensory stimulation. The individual may then behave in a challenging way in an attempt to meet unmet needs. Kovach et al. (2004) further indicated that agitation is produced by both high and low stimuli situations (imbalances in sensoristasis). Algase et al. (1996) discussed how lack of fulfillment of needs in individuals with dementia compromises their behavior. That is, behavioral symptoms of dementia stem from an individual's thwarted need or goal. An early variant of this model was espoused by Hall and Buckwalter (1987), who suggested environmental vulnerability/ reduced stress-threshold causes the individual's challenging behaviors.

One other important theory involves the behavioral or learning model, where reinforcement contingencies are evaluated as the way to explain behaviors (Teri, McKenzie, & LaFazia, 2005). Teri et al. argue that an applied behavioral analysis is sufficient for change in residents with psychiatric problems (Teri et al., 2000). This method applied in a validating environment with a motivated staff has merit. One variant involves the STAR method (Teri et al., 2003), an A-B-C analysis (i.e., antecedent–behavior–consequences) of a situation providing an understanding of the conditions that merit change. Again, as behavior is determined by a specific antecedent and maintained by a specific consequence, trained staff can apply behavioral principles.

In recent years, other models have been proffered. Several authors have advocated holistic models with its many factors of symptom presentation (Kitwood, 1995; Kitwood & Benson, 1995) and those associated with disruptive behaviors such as agitation (Volicer & Hurley, 2003). Several writers highlight the use of continuing abilities. For example, use of habitual skills, humor, emotional memory or emotional awareness, sociability, sensory appreciation, motor function, music responsiveness, and long-term memory have been promulgated (Bowlby-Sifton, 2000).

Validation by the caregiver for the declining resident was operationalized (Feil, 1999). Generally, people with dementia attempt to make sense of their changing lives. Many attempt to resolve earlier developmental and other issues. Zgola (1999) noted that caregivers can be more successful by effectively using their relationship with the resident. In fact, the relationship is considered the centerpiece of care, without which there is no positive environment. She holds that good care is dialogical and can be fulfilling for the caregiver.

Finally, there are those who consider cognitive rehabilitation as a treatment option for LTC residents (e.g. Camp, 2000). West, Welch, and Yassida (2000) describe the results of seminal studies on cognitive rehabilitation regarding older people. They explained that, memory strategies, even complex techniques, can be learned by older adults; training can lead to improvement that can last for long periods; and, training tends to be task-specific, with transfer of tasks incorporated into the training regimen.

In a much cited metaanalysis of this area, Verhaegen, Marcoen, and Goossens (1992) endorse memory training and its positive effects. This group noted that four factors enhance training: pretraining, group sessions, shorter sessions, and younger participants. Hyer and Rebock (2005) also espouse a "mild" rehabilitation approach involving the patient and caregiver dyad. The application of memory retraining using a problem-solving paradigm, where supports are introduced over time, is a proactive approach. Use of effective communication, validating emotions, functional assistance, maintaining social skills, decreased environmentally-based stimuli that can contribute to confusion, and a constant effort to understand the person are also used.

Relevant Medical Issues in LTC

Modal Medical Problems in LTC. In LTC facilities Medicare or Medicaid certified nursing homes are required to complete a comprehensive assessment of each resident's functional capabilities, diagnoses, and interim complications in the MDS. The top disease diagnoses reported in the MDS Active Resident Information Report (Centers for Medicare and Medicaid Studies, 2004) include hypertension, depression, diseases of mood or behavior, and dementia. Most of these involve psychiatric problems.

Continuity of Care and Functional Status. The discontinuity in care between an acute care hospital and an LTC setting is considerable (McGilton, 2002). The health care provider team is changed upon LTC admission. Mold, Fryer, and Roberts (2004) present several benefits of continuity of care, including improved patient, physician, and staff satisfaction, increased trust of physician, better identification of medical problems, better coordination of care, more preventative services, decreased cost of care and greater likelihood of patient disclosure of emotional problems. These benefits are particularly relevant for older adults who are likely to have several comorbid diseases, increased frailty, and require careful management of multiple medications.

New residents arrive at LTC facilities with a "life" behind them. As such, they are likely to come with poorly defined premorbid information, both physical heath and incomplete narratives. There could be mental health problems that may or may not have been diagnosed, acknowledged, and treated prior to the onset of change that brought them to the facility. Conventional wisdom that a caregiver will be relieved of a great burden when a family member is moved to a nursing home has been challenged (see Schulz et al., 2004). For example, a daughter who took care of her mother now has to negotiate for her parent's care at the facility, and worry about her care in new ways.

People enter LTC because of some inability to function independently (Williams, 1998). Functional status can be monitored, as a way to assess treatment responses and provide important information when evaluating a care plan (Arseven, Chang, Arseven, & Emanuel, 2005). Functional assessment is typically classified into basic and instrumental activities of daily living (ADLs and IADLs respectively). A person who can carry out basic ADLs functions independently; inability to perform any of these activities necessitates some assistance. The amount of assistance necessary is a function of level of dependence (Heath, Gartenberg, & Beagin, 2005).

Pain. In nursing home settings where a majority of residents suffer from dementia, 67% of the residents reported pain, but only 37% had pain recorded in the record (Ferrell, Ferrell, & Rivera, 1995). Pain is viewed through many different lenses, including cultural, past health experiences, and religious perspectives. Currently, the goal is to individualize the tolerance, of all forms of physical discomfort, with the needs for joint mobility, required skin wound care, or other specific rehabilitation measures that are often part of the daily routine within LTC rehabilitation units (Feldt, 2004). Pain issues and less-than-optimal motivation for participation in rehabilitation are often prime triggers for the enlisting of mental health care providers (Snow et al., 2005).

Typology of Long-Term Care Facility Residents

Resident Populations. A useful typology to consider in LTC involves the interaction between the mental and medical health needs of residents that is based on the "trajectory" of the anticipated course of care. Heath et al. (2005) identified three groups of nursing home residents, namely actively rehabilitating, actively dying, or surviving. The "actively rehabbing" resident justifies the role that LTC facilities play within the broader health care system in providing a site for rehabilitation following an acute illness. Older adults are generally admitted to a subacute rehabilitation unit within an LTC facility. Depending on the functional status and available resources, follow-up home or community-based rehabilitation may be recommended after discharge.

In general, many of the mental health issues confronted by these "rehabbing" residents of LTC facilities involve the practice of behavioral medicine (BM). In fact, these residents are more similar to patients in a hospital and usually will be in the nursing facility for less than 60 days.

Common medical disorders in LTC facilities include cardiovascular disorders, diabetes, sleep disorders, multiple sclerosis, chronic pain, COPD, and cancer. The multidisciplinary approach utilizing BM techniques includes psycho-education, self-management, treatment adherence, and monitoring outcomes (Casciani, 2003).

The role of physicians, central to much of the direct acute care provided in the hospitals, is diminished in LTC settings though physician involvement in the treatment of LTC rehabilitation residents is still important in the monitoring of chronic conditions that may influence the trajectory of the rehabilitation process. However, after helping to establish the priorities and approach to care and services, physician care is largely supplanted by nursing and rehabilitative therapists who deliver such services as wound care, medications and the physical, occupational, speech, and swallowing training, and other interventions that may be prescribed as part of the rehab process. Many of these interventions are intended to require greater participation by the individual resident to demonstrate recovery of function and health.

A smaller proportion of residents are in the end-stages of life. While obviously many older residents within LTC are nearing the end of their natural life, the health care perspective for this particular group of residents, classified here as "actively dying," is different. They are formally recognized as no longer "living with" their conditions but rather now "dying from" them and have a prognosis of 6 months or less left to live. The most commonly cited diagnoses among these residents include cancers where interventions focused on cure are no longer considered effective (Buchanan, Choi, Wang, & Ju, 2004). Recent efforts have sought to better define the end stage of noncancer conditions such as advanced AIDS, ischemic heart diseases, or progressive neurological conditions. Hospice often takes over the care of the dying resident. In 2000, about 60% of newly admitted LTC facility residents whose MDS indicated a limited prognosis were formally under a hospice organization's care (Miller, Mor, & Teno, 2003).

Many LTC residents who are "actively dying," then, will not be under the formal care of a hospice organization during their terminal care. Mental health issues that arise among such residents whose care is explicitly centered on terminal care is first and foremost, one of relief of pain and discomfort, along with its associated depression and despondency (Miller et al., 2003). The dying process is highly individualized (personal) and will often incorporate the attitudes and approaches of those surrounding the dying individual, including the professionals, family, caregivers, and friends.

Challenging issues that also arise within the mental health care of such individuals can stem from organic brain lesions associated with the terminal illness. Structural brain lesions can produce unique and often quite disabling behavioral manifestations. Hallucinatory or delusional perceptions can also be associated with delirium episodes related to metabolic and endocrinologic effects of some cancers.

While the occurrence of these rather dramatic events are evident to all discussed typologies, other more subtle "red flag" triggers in this population of LTC residents include isolation and withdrawal. The psychologist is singularly adept at assessing mediating factors to treat the "person of the dying resident." While there may be no realistic expectation for a cure of the disease, hope can come from other sources, including, comfort, dignity, intimacy, and salvation, for example.

The largest proportion of LTC residents include the "surviving." Heath and colleagues (1995) identify this group to include residents in various stages of dementias, such as Alzheimer's disease. Slowly progressive dementias may cause functional impairment that requires long-term care, but has not yet led to immediate threats to continued survival. Such dementing illnesses may either be explicitly identified as the primary diagnosis for the individual or be listed among a variety of chronic conditions that have accumulated to an extent that institutional placement was required. Often the coexistence of dementia with depression further complicates the therapeutic prioritization (Burrows, Satlin, Salzman, Nobel, & Lipsitz, 1995).

Effective psychological interventions (i.e., behavioral, cognitive or dynamic) include assessment, planning, and treatment (see previous chapters of this text). In an LTC setting, the resident seeks equilibrium, a best fit interaction between self and environment. An adaptive balance between security and autonomy is naturally sought.

One model (implied from the above models), SOC (selective, optimization, and compensation; Baltes & Baltes, 1990), has defined how this process can be best accounted for and altered. That is, how the declining resident in an LTC setting can select, optimize, and compensate for debility, outcomes critical for quality of life. The psychologist's task is to alter the system or the environment or both, and empower the resident to respond maximally in this setting. In effect, the psychologist is asked to minimize the impact of age- or disease-related losses while maximizing potential gains in their environment. The psychotherapeutic approach, relating to the mental health needs of "surviving" residents includes confronting both learned helplessness and the attitude that depression or ineffective functioning is inevitable. The discussion below focuses on this population.

Psychiatric Treatment in LTC

It cannot be overemphasized that the identification and treatment of mental disorders in an LTC setting must be done in partnership across and among professional disciplines (e.g., medicine, psychiatry, psychology, social work, nursing, pharmacy, and administration). An expert panel (American Geriatrics Association and American Association for Geriatric Psychiatry, 2003) published a consensus statement on "Quality Mental Health Care in Nursing Homes." The identified principles can also apply to assisted living communities.

The panel established a number of overarching principles to guide the improvement of policy and practice in caring for residents' mental health needs, reinforcing the idea that the living environment should be structured and operated in such a way as to promote the mental health of the resident population. These principles assert (1) residents should be allowed to maintain an autonomous and pleasant lifestyle and be treated with respect and dignity; (2) quality mental health care relies on the presence of high-quality medical care; (3) State and Federal regulatory efforts must focus on treatment and assessment; (4) appropriately qualified mental health specialty providers must be adequately reimbursed to assess and treat residents with mental health problems; (5) residents should be encouraged to develop trusting relationships, and want to feel supported, and productive, while retaining as much autonomy as possible. These older adults want to decide when to wake up, take care of daily chores, or whether to be alone or participate in the day's group activities; (6) nursing homes must maintain adequate and permanent staffing in order to personalize and strengthen staff–resident relationships; and (7) an environment that is less institutional and more homelike with the presence of pets, plants, and children promotes a higher quality of life and aids the management of resident depression and behavioral symptoms.

Treatment of Mental Illness. Treatment is directed at the dominant symptoms or most problematic behavior or concerns. Interventions are aimed at minimizing the adverse impact of the milieu or environmental factors. Realistic goals focus on reducing the target symptoms, not completely eliminating them (Sutor, Rummans, & Glenn, 2001).

OBRA 1987 (Omnibus Budget Reconciliation Act, 1990), enacted in 1990, provided guidelines for use of psychotropic medications and chemical–physical restraints. OBRA guidelines were identified when antipsychotics must *not* be prescribed. These include wandering, restlessness, anxiety, poor self-care, insomnia, depression without psychosis, uncooperative behavior, agitation with no harm to self or others, and impaired memory.

Acceptable indications for antipsychotic use in nursing homes, according to OBRA guidelines include schizophrenia, Huntington's disease, schizo-affective disorder, psychotic and delusional disorders, Tourette's syndrome, mood disorders with psychotic features, and organic mental disorders (dementia and delirium) with associated psychotic or agitated behaviors, and with appropriate justification and documentation. Documentation attests that the target behavior is permanent or persisting and causing a danger to patient or others or severe functional impairment.

Routine and not "PRN" orders are preferred. There should be an attempt to reduce the medicine's dosage twice a year. Documented justification to continue the medicine includes a reduction in dosage that failed, and that with the medicine, the symptoms are stable and without side effects. There should be ongoing monitoring for side effects, particularly tardive dyskinesia, cognitive or behavioral impairment, hypotension, and parkinsonism.

The impact of OBRA on psychotropic drug prescribing reveals an overall decrease in the rates of antipsychotic use, increase in antidepressant use, decrease in hypnotic use, and relatively little change in anxiolytic use (Ryan, Kidder, Daiello, & Tariot, 2002). Since the enactment of OBRA, however, there is scant evidence to indicate that the use of nonpharmacological interventions has increased (Cody, Beck, & Svarstad, 2002). Challenges to the use of nonpharmacological interventions include, in part, inadequate communication between the medical director, physicians, and facility staff particularly concerning the residents' mental health needs. Additionally, these interventions require training and time, a coordinated team.

Depression. Depression reduces quality of life and exacerbates problems, including functional deficits (Nanna, Lichtenberg, Buda-Abela, & Barth, 1997), behavioral disturbances (Ruckdeschel, Thompson, Datto, Streim, & Katz, 2004), suboptimal nutrition (Mitnitski, Song, & Rockwood, 2004), nonadherence with treatment plans (Murphy & Alexopoulos, 2004), pain (Parmalee, Katz, & Lawton,

1991), and disability, as found in dementia (Blazer, 2003). Depression also contributes to excessive morbidity (Parmelee, Katz, & Lawton, 1992) and mortality (Cohen, Hyland, & Kimhy, 2003).

A thorough medical assessment should be included in the assessment of new onset or worsening of depression. This problem is modal in LTC both as a syndrome and at a subclinical level. Medically, this exam should address pain assessment, nutritional status, worsening of chronic medical conditions, recent onset of a new medical condition, and the administration of medications that can possibly alter mood or cognition. Unless recent results are available, one must consider laboratory and diagnostic testing as determined by the findings of the history and physical examination. Testing may include a complete blood count, thyroid function, electrolytes, including serum calcium and magnesium, Vitamin B12 level, and urinalysis (for evidence of a urinary tract infection). The guidelines also recommend an evaluation for the presence of suicidal thoughts or psychosis.

While provision of psychological services is an appropriate, perhaps even preferred first-line approach to depression, the first intervention is often a medication trial. If, after 6 or more weeks of treatment there is failure to adequately respond, a referral to a mental health professional is clearly in order. Again, there is evidence to support a psychologically based approach as the first line of treatment to minimize or prevent the problems associated with adverse medication reactions and polypharmacy (see Arnold, in press, for a discussion of polypharmacy and older adults).

The Clinical Practice Guidelines (CPG) Project, initiated by the American Medical Directors Association (AMDA; American Medical Directors Association, 1996) offers another recommended approach to assessment and treatment of depression in LTC settings. The AMDA guidelines state that an interdisciplinary assessment is crucial to identifying signs and symptoms of depression. Further, they recommend the use of scales such as the Geriatric Depression scale (GDS), Cornell Scale for Depression in Dementia, or the Center for Epidemiologic Studies of Depression Scale (CES-D; Radloff, 1977). These measures were also reviewed and recommended by Hyer, Carpenter, Bishman, and Wu (2005).

In addition, significant information can be garnered from other sources, such as rehabilitation and recreation personnel. Importantly, the AMDA guidelines recommend consideration of a consultation from a geriatric psychiatrist when there is uncertainty regarding the diagnosis of depression or when severe, urgent, or emergent situations arise. This is especially important in the presence of suicidal ideation, homicidal ideation, refusal to eat or drink because of depression, or the presence of psychotic symptoms. Further, medication must always be ruled out as a contributing etiologic factor. The use of levodopa-carbidopa, clonidine, b-adrenergic blockers, benzodiazepines, and barbiturates especially, are among the medications that are known to cause depression as a side effect.

Data suggest efficacy of selective serotonin reuptake inhibitors (SSRIs) in cognitively intact patients, but have limited efficacy in those with severe dementia and depression (Katz, 1999; Schneider, Dagerman, Insel 2006). Of note, the authors found no studies that combine antidepressants with nonpharmacological interventions. According to published expert consensus guidelines for the treatment of depression in older adults (Alexopoulos, Katz, Reynolds, Carpenter, & Docherty, 2001), SSRIs are considered the first choice among antidepressants. Given a low probability of drug–drug interactions, the guidelines cite sertraline and citalopram as mentioned first-line agents.

For any medication intervention with the geriatric population the rule "start low and go slow" must be followed, as older adults often have comorbid medical illnesses and take other medications. These factors increase the risk of untoward side effects. There are no data on the use of serotonergic–noradrenergic agents (e.g., Venlafaxine, Effexor) in the nursing home, and yet they are often a second-line option for the treatment of geriatric depression because they are generally considered to be relatively well tolerated.

Underlying vulnerabilities to the potential side effects of medication must be kept in mind while selecting a particular antidepressant agent. For example, a patient with chronic constipation, in addition to depression, can be given a trial of an agent that is often associated with causing loose bowel movements, such as sertraline (e.g., Zoloft). Adverse effects of the SSRIs include gastrointestinal symptoms, such as nausea, emesis, and diarrhea. In addition, these agents reportedly can also cause extrapyramidal symptoms and hyponatremia.

Mirtazapine (e.g., Remeron), a serotonergic agent with a mechanism of action distinct from the SSRIs, is associated with sedation and weight gain, and so it is often used with older adults when loss of appetite and insomnia are the predominant symptoms of depression.

Bupropion (e.g., Wellbutrin), a nonserotonergic compound, is contraindicated in individuals with seizure disorder as it lowers seizure threshold. Individuals with very severe depression, severe weight loss, persistent and active suicidal ideation and depression unresponsive to pharmacologic intervention may require electro-convulsive therapy (Peskin, 2003).

Per other consensus guidelines (AGS & AAGP, 2003), patients who meet criteria for minor depression should be treated with nonpharmacological interventions and observation of response for up to 2 months. Depending upon symptom severity, these patients may need antidepressant treatment. Patients who meet criteria for major depression with psychotic features should be carefully evaluated and monitored when antidepressant and antipsychotic medications are indicated.

Dementia. There are two issues with dementia for psychiatry, namely cognition and psychiatric symptoms. We first address cognition. At present the pharmacological treatments for dementia of the Alzheimer's type (DAT) approved in the United States involve ChEIs (cholinesterase inhibitors – Donepezil, Galantamine, Rivastigmine) and the newer Memantine (e.g., Namenda), an *N*-methyl-d-aspartate receptor antagonist. To date, data indicate that ChEI (cholinesterase inhibitors) use in the patients with DAT will benefit the person by enabling an increase on cognitive scales, and that this improvement delays decline for periods of up to two years (Anand, Hartman, Sohn, Danyluk, & Graham, 2003).

Indeed, recent reviews of the ChEIs indicate that these medications should be used early in the disease process and continued for a longer period of time than was originally thought (see Hyer & Ragan, 2002). As a result, it is increasingly likely that ChEIs will be used with growing frequency in medical settings, especially LTC facilities. ChEIs are thought to engage the explicit or effortful memory system, enhancing the cholinergic system in the brain, which is most related to new learning, short-term memory, executive functioning, and recall. These are the areas initially damaged during DAT (Grossberg & Dasai, 2003). Outcomes, therefore, based on the use of these agents have involved measures of declarative memory, such as the cognitive subscale of the Alzheimer's Disease Assessment Scale (ADAS Cog) or the Mini-Mental State Exam (MMSE).

There are important limitations, however. Results of a recent study out of England on the long-term use of Aricept provide an example. Courtney et al. (2004) studied 565 people with mild to moderate DAT. They were assigned to Aricept or a placebo condition and were followed over 3 years. Although those taking the active drug had slightly higher scores than the placebo group, there were no differences in their behavioral and psychological symptoms, or in the emotional well being of the people taking care of the patients.

Recent reviews of the efficacy, safety, and tolerability of the ChEIs have also raised concerns (Anand et al., 2003; Sink, Holden, & Yaffe, 2005). Subjects in clinical trials appear not to be representative of routine clinical populations in terms of socioeconomic, education, race, and health and behavioral status. Such complications have also been found in older populations recruited for clinical trials using atypical neuroleptics. Schneider et al. (2006) showed that only 4.4% to 7.9% of subjects in a large clinic database would have met entry criteria for a clinical trial. In a review, Clegg et al. (2002) determined that improvements in global health and/or cognition when taking donepezil (e.g., Aricept), rivastigmine (e.g., Exelon), or galantamine (e.g., Reminyl) were "small and not clinically significant."

Dementing patients are suffused with psychiatric symptoms. ChEIs, especially tacrine, donepezil, rivastigmine, and galantamine, have been studied quite extensively in community settings for their efficacy on behavior and functioning in Alzheimer's dementia, as well as cognition. Tariot and colleagues (Tariot et al, 2001) reported on a 24-week double-blind placebo-controlled study of donepezil, with 208 nursing home patients who had Alzheimer's disease or Alzheimer's disease with cerebrovascular disease. The donepezil-treated group showed improved agitation and aggression as noted on the Neuropsychiatric Inventory – Nursing Home Version scale, as compared with the placebo-treated group. Agitation and aggression were the most common behavioral symptoms associated with Alzheimer's disease in this study. Of the 64% of patients who displayed these behaviors, 67% were noted to experience symptom relief with the use of donepezil.

In general, ChEIs appear to be mildly if at all effective with psychiatric symptoms of dementia. In a review of pharmacological treatment of neuropsychiatric symptoms of dementia, Sink et al. (2005) conclude that extant medications are not particularly effective for the management of neuropsychiatric symptoms of dementia.

Two metaanalyses (Olin & Schneider, 2002; Trinh, Hoblyn, Moharty, & Yaffe, 2003) and six additional randomized clinical trials (RCTs; Courtney et al., 2004;

Erkinjuntti et al., 2002; McKeith et al., 2000; Tariot et al., 2001) of various ChEIs for those with neuropsychiatric symptoms have been published. Five of the eight studies report statistically significant benefit on neuropsychiatric symptoms. However generally, the magnitude of the effect is small and of questionable significance (Sink et al., 2005). These findings also apply to Memantine (Reisberg et al., 2003; Tariot et al., 2004), where results indicated mild changes that are not clinically significant.

Importantly, additional psychosocial interventions have not been systematically applied as add-on in these studies. Further, there is a special gap in understanding when it comes to the involvement of caregivers, the key persons in the treatment of older adults with dementia. Recently, for example, Kaufer, Boroson, and Sadik (2005) reported a positive difference in caregiver time after use of a ChEI. One implication is that there is no substitute for the careful and thoughtful involvement of caregivers in this population, especially when it comes to issues of memory.

Use of Antipsychotic Medications. Use of atypical antipsychotic medications for psychiatric symptoms in dementia is often the treatment of first choice. Education of staff to recognize, assess, treat, and monitor behavioral symptoms is essential. The MDS used in nursing homes is inadequate for this task. It is important to consistently identify and quantify the specific behaviors of concern, such as aggression and verbal agitation or restlessness. Precipitants to the behaviors must also be identified. Verbal and nonverbal communication of unmet needs, such as hunger, thirst, the need to use the bathroom, for warmth, sleep, and touch also require assessment.

New onset of symptoms or changes in clinical presentation should prompt assessment for psychosis, depression, anxiety, insomnia, drug effects, and prescription drugs overuse or withdrawal, especially in the presence of medications like benzodiazepines. Assessment for hearing and visual problems must be considered when there is a possibility of sensory deprivation.

Again, ruling-out possible medical or pharmacological contributions or causes to signs and symptoms is necessary. New onset or changes in symptoms should prompt a vital sign check (blood pressure, pulse, respirations, and temperature) and evaluation for infections, dehydration, pain or discomfort, delirium, fecal impaction, and injury. An estimated one-third of the older adult population develops psychiatric symptoms secondary to underlying medical disorders (Sutor et al, 2001).

Medical causes contributing to behavior symptoms can come from a wide array of conditions, including dehydration, congestive heart failure, diabetes, hypothyroidism or hyperthyroidism, urinary tract infection, pneumonia, and cellulitis. Neurological conditions such as stroke, subdural hematoma, and traumatic brain injury can also contribute to psychiatric symptoms. In addition, drug effects can be a potential factor in causing anticholinergic induced delirium. The presence of delirium in LTC (acute or chronic) is highly prevalent and the health care provider must be ever vigilant for its presence and its etiology.

The available evidence suggests that atypical neuroleptics are better tolerated than conventional neuroleptics and more efficacious than placebo, in decreasing agitation in patients with dementia. Recent reports reveal that these agents may increase the risk of glucose intolerance and adverse cerebrovascular events. Clozapine is occasionally recommended for treatment of refractory older adults with movement disorders, such as parkinsonism or tardive dyskinesia and psychosis. Clozapine use can significantly increase the risk of agranulocytosis. Regular standardized blood monitoring is required (Kasckow, Mulchahey, & Mohamed, 2004).

In April 2005, the United States Food and Drug Administration (FDA) requested that all manufacturers of atypical antipsychotic medications add a black box warning to their prescribing information regarding prescribing these medications for older adults with dementia-related psychosis (FDA, 2005). According to the FDA, analyses of 17 placebo-controlled trials that enrolled 5,106 older adult patients with dementia-related behavior disturbance revealed a risk of death of 1.6–1.7 times in the drug-treated patients than that seen in placebo-treated patients.

Clinical trials were performed with Zyprexa, Abilify, Risperdal, and Seroquel. Over the course of these trials, averaging about 10& weeks, the rate of death in drug-treated patients was about 4.5%, compared with a rate of 2.6% in the placebo group. The causes of death varied, appearing to be either cardiovascular or infectious. The black box warning further states that the FDA has not approved Abilify, Seroquel, Geodon, Risperdal, or Zyprexa for the treatment of patients with dementia-related psychosis.

Given the black box warning, which includes increased cardiovascular risks (stroke), metabolic side effects such as weight gain, hyperlipidemia, insulin resistance, and diabetes, a careful approach to the treatment of dementia-related psychosis is warranted. One must identify target symptoms (psychosis), and only in the presence of dangerousness to self or others or impairment of functioning (resistance to care due to psychotic symptoms) is the use of atypical antipsychotic medications justifiable. The responsibility of clinicians to educate the staff and family members of potential side effects also cannot be overstated. In addition, nonpharmacological interventions should always be used extensively.

Currently, atypical antipsychotics are the only medications that are effective for dementia-related psychosis. They will continue to be used in specific situations where nonpharmacological interventions, anxiolytics, and antidepressants have been exhausted. Again, it is important to underscore the compelling research concerning the efficacy of psychosocial interventions.

Anticonvulsant Medications. In addition to the emergence of atypical antipsychotic agents as commonly used agents for dementia-associated behavioral problems, the use of anticonvulsant medications, such as carbamazepine (e.g., Tegretol) and sodium valproate (e.g., Depakote), is now common practice. A placebo-controlled study, using an average dose of 300& mg& day^{-1} of carbamazepine, showed that the drug was well tolerated by patients and significantly decreased aggression and agitation (Tariot et al, 1998). Potential side effects of carbamazepine include drowsiness, gastrointestinal distress, ataxia, rash, elevated hepatic enzymes, and drug interactions with other agents. More seriously, this agent has also been associated with aplastic anemia and agranulocytosis (a life-threatening event).

Sodium valproate has reportedly been effective in the treatment of behavioral disturbances in dementia and with fewer drug interactions than carbamazepine. Porsteinsson, Tariot, and Erb (2001) reported on the use of sodium valproate in a randomized, placebo-controlled, 6-week study of 56 nursing home patients with dementia and agitation. The average dose of divalproex sodium used was 840 mg day^{-1}. The authors noted that their results suggested improvement in the agitation subscale of the Brief Psychiatric Rating Scale. Potential side effects of this agent include nausea, vomiting, sedation, diarrhea, ataxia, and tremor. Hepatotoxicity and pancreatitis have been noted to be rare, but serious, complications of the use of sodium valproate. When this medication is administered, complete blood counts, hepatic enzymes, and serum drug concentrations need to be monitored regularly to prevent toxicity. Gabapentin, lamotrigine, and topiramate are newer anticonvulsants whose efficacy and tolerability have not been systematically evaluated for the treatment of dementia-associated behavioral changes.

Anxiolytics. Benzodiazepines, such as lorazepam and oxazepam, are often used in the management of agitation, but their efficacy has not been comprehensively studied. These agents are generally safe when used in low doses, but with increasing doses, common side effects that emerge include oversedation, ataxia, confusion, and paradoxical agitation. In addition and importantly, with long-term use, tolerance and dependence are likely. Zolpidem, a hypnotic, was shown to reduce nighttime wandering in two patients (Shelton & Hocking, 1997).

The antianxiety agent, buspirone, has shown efficacy in case reports (Sakauye, Camp, & Ford, 1993) in reducing anxiety, aggression, and agitation in patients with dementia. The drug is generally well tolerated. However, if used in combination with SSRIs, the patient is at increased risk for the development of serotonin syndrome (confusion, tremors, hyperthermia, hypertension, and seizure).

As with the treatment of depression, expert consensus guidelines have been published regarding the use of medications for the treatment of agitation in older patients with dementia (Expert Consensus Guideline, 1998). The guidelines recommend initiating therapy with risperidone at low doses of 0.25–0.5mg day^{-1}, with an average maximum dose of 1–1.5 mg day^{-1}, if there is evidence of psychosis. These guidelines also recommend the use of environmental interventions in addition to medication treatment. If agitation is due to depression and/or anxiety, an SSRI trial is also recommended. If insomnia is a contributing factor, trazodone can be prescribed. Combination medication therapy is indicated only after two different trials with two different classes of agents have proven unsuccessful.

Other consensus guidelines (AGS & AAGP, 2003) recommend that nonpharmacological interventions are indicated, rather than medications, if there is no threat to the resident or others, medical conditions are being addressed, and there are no evident psychotic symptoms. Atypical antipsychotics are the first-line agents of choice for severe behavioral symptoms with psychosis. When medications are used to manage behavior disturbances again, one must document the specific target symptoms being addressed. There is no consensus on the choice of medication for nonpsychotic behavioral symptoms. Further, there are no comparison studies between pharmacological and nonpharmacological interventions.

Psychology: Integration

Problems. There is agreement that the goal of care for a resident is to maintain function, and where possible, to increase quality and quantity of life, even in a long-term care environment. The interface among medicine, psychiatry, and psychology is not an easy task but it is essential for true interdisciplinary team functioning. There is no gold standard, and there are road blocks and competing professional issues, all approximations of best practices.

It is the current reality in long-term care settings that medication is the default rule of care. Variance of the effect of the drug resides mostly in the individual and not in the compound. The general approach to pharmacotherapy, however, is to alter behavior problems – effective change is marked and continued, and ineffective change is noted and other medications tried. However, in light of the above evidence concerning the limitations and risks of many psychotropic medications, their use as a first-line approach or as a single approach can be an incomplete and problematic approach to care.

When psychotropics are necessary, they must be justified, monitored regularly, and, considered for tapering. Optimal medical psychiatric care, however, is often not applied in practice. As a consequence, a kind of "medicament thinking" ensues. In such a scenario, staff input is biased in the direction of medication: the target behavior identified, medication is prescribed, and a medication watch ensues. All change is associated with the medication. This includes a careful monitoring of side effects, which are considered eventful as the medication is "working." If there are problems, it is viewed that the wrong drug was chosen, dosage was too high or too low, or duration of trial was inadequate. In effect, all attributions of change or lack thereof are considered due to the medication. Often this bias shuts down or at the least alters dialogue about psychological and social intervention with their lowered risk of untoward consequences.

At present, there are no double-blind, placebo-controlled, RCTs in LTC. Extant trials in LTC do not meet established criteria for empirically supported treatments (Chambless & Ollendick, 2001; Kazdin, 1994), nor have they undergone the scrutiny normally applied to criteria of randomized trials (Foa & Meadows, 1997). Hyer and colleagues (Hyer et al., 2005) argued for a relaxation of RCT standards in these settings. There is reason to assert the transferability of cognitive behavior therapy (CBT), for example, to LTC (Snow, Powers, & Lyles, in press) and CBT should be considered early on in the treatment process. Several models are currently under evaluation (e.g., Hyer, 2004). Cohen-Mansfield (2001) reviewed 83 studies on psychosocial treatments and suggested that combined and individualized therapies can be applied in LTC settings. The research on the practice of psychology in LTC then is necessarily less exact than medication usage, but nevertheless provides strong evidence that psychological (i.e., cognitive, behavioral, dynamic) interventions work.

Assessment. Assessment remains a psychological staple in LTC settings (see Chap.3 of this volume). Assessments that should be part of general evaluations for residents in long-term care settings include signs of cognitive decline, extent of depression, and modal psychopathologies that are clearly interfering with the resident's experience and/or functioning.

The most important measures in LTC involve cognition. This variable predicts varied outcomes better than just about any other variable. Knowledge of mental status then provides important data on residents – trajectory of decline, probable problems, needed environmental supports, and information on best match for interventions. This is the domain of psychology.

Psychological Treatment. In their review of treatment studies in LTC, Hyer et al. (2005) found sufficient support for the judicious application of CBT and reminiscence therapy for most residents, as well as behavior modification in the context of dementia. Increasingly too, there are treatment manuals being developed for this population. The appropriate admixture of medication and psychosocial treatments is clearly necessary in LTC.

The clinical task for any resident is to: (1) establish and articulate a clear problem list and/or diagnosis, (2) educate the resident and caregiver about prognosis and stages of disease, (3) discuss the resident's present state and treatment options, (4) assemble comprehensive treatment plan (medical, social, financial, legal), (5) establish caregiver support and sense of partnership, and (6) confirm ongoing support throughout the entire course of illness. Clinicians often affirm that they act as an advocate for the resident. In clinical staff meetings, discussion usually involves the medical problem, medication effects, cognitive impairment, mental syndrome, as well as history and causes, and environment or change in environment.

The Challenges. Psychologists face several professional challenges in trying to foster integration in LTC. First, they must be informed about the milieu and professional practice in LTC. Knowledge of frail elders in LTC is strikingly different from that of other residents. In general, there is a lack of awareness of the efficacy and application of psychological and social models. This is critical in any setting but especially so in LTC where psychologists have the training to assist or lead in the input of care. Psychologists must also be mindful of the other professional staff. At present, medical and nursing care can reasonably address limits in several areas of functioning, including pain, sensory problems, sleep problems, and limits on autonomy. To function well in an LTC setting there is a need especially for an integrated and multidisciplinary team (i.e., interdisciplinary team functioning). This involves a shared responsibility, recognizing the strengths of other professions, with a high degree of respect for the health care recipient and absent the specialist mentality; thus, enforcing the idea that the "team as a whole" takes responsibility for program effectiveness. Leadership functions are shared and there is no hierarchical team organization, propagating an atmosphere of creative experimentation.

LTC teams across the country attempt to manage people by reducing problem behaviors. At its base, this involves discovering the meaning of behavior, considering environmental interventions, providing socialization, and developing behavioral strategies (Grossberg & Desai, 2003) among others.

As noted earlier, psychological and social interventions should be implemented as a first-line approach. Unfortunately, the practice in most settings is medication and containment of the "problem" (Cohen-Mansfield, 2001). Psychologists, along with those who prescribe medication (and in some areas this may be one and the same person) need to integrate care for the best mix of medication and psychosocial approaches. The practicing psychologist in LTC then must be clinically informed, a true team member (or leader), and trainer or educator. In addition, the psychologist must be willing to patiently design and direct individually tailored interventions.

Conclusion

The challenge for psychology, in the treatment of residents in LTC, is to strike a balance between fighting the disease and coming to terms with limitations. This is a rather difficult undertaking as values become part of the care process and data are wanting. Who can say that a resident is better off who remains in bed most or all of the day and is "medicated" but less agitated? Who can say whether the chemical management of a confused resident is not appropriate in terms of staff time? Additionally, LTC facilities fight for better care in a sea of regulations. They attempt to apply the "OBRA path," but are not able to fulfill this mission due to inherent limits – staff shortage, ignorance, or turnover, use of professional consultation, patience, and a host of other system variables. The practice of mental health in LTC is too often dictated by payment systems and by regulations, often to the detriment of the resident.

The integration of psychology with the other professions is important and is especially so with psychiatry, as these are the two disciplines most involved in mental health care. These collaborations can make an enormous difference in the overall quality of care and services provided to long-term care residents. These disciplines can apply the technologies of medication and/or psychological or behavioral and psychosocial interventions carefully and deliberatively. A thoughtful dialogue, administrative support, and knowledge of the environment and resident can improve outcomes of care. Like emergent principles in any environment, the LTC setting is more than the sum of its parts – administration, residents, rules, professions. In this world, psychology can make a difference.

References

Alexopoulos, G. S., Katz, I. R., Reynolds, C. F., Carpenter, D., Docherty, J. P. (October 2001). *The expert consensus guidelines: Pharmacotherapy of depressive disorders in older patients.* A Postgraduate Medicine Special Report. New York: The McGraw-Hill Companies. http://www.psychguides.com/ecgs5.php

Algase, D. L., Beck, C., Kolanowski, A., Whall, A., Berent, S., Richards, K., et al. (1996). Need-driven dementia-compromised behavior: An alternative view of disruptive behavior. *American Journal of Alzheimer's Disease, 11*(6), 10-19.

American Geriatrics Society and American Association for Geriatric Psychiatry. (2003). Consensus statement on improving the quality of mental health care in U.S. nursing homes:

Management of depression and behavioral symptoms associated with dementia. *Journal of the American Geriatrics Society, 51*, 1287–1298.

American Medical Directors Association. (1996). *Depression clinical practice guideline.* Columbia, MD: American Medical Directors Association.

Anand, R., Hartman, R., Sohn, H., Danyluk, J., & Graham, S. M. (2003). Impact of study design and patient population on outcomes from cholinesterase inhibitor trials. *American Journal of Geriatric Psychiatry, 11*(2), 160-168.

Arnold, M. (in press). Polypharmacy and older adults: A role for psychology and psychologists. *Professional Psychology Research and Practice*

Arseven, A., Chang, C. H., Arseven, O. K., & Emanuel, L. L. (2005). Assessment-instruments. *Clinics in Geriatric Medicine, 21*(1), 121-146.

Baltes, P. B., & Baltes, M. M. (1990). Psychological perspectives on successful aging: The model of selective optimization with compensation. In P. B. Baltes & M. M. Baltes (Eds.), *Successful aging: Perspective from the behavioral sciences* (pp. 1-34). Cambridge, England: University Press.

Bartels, S. J., Maok, G. S., & Dums, A. R. (2002). Models of mental health services in nursing homes: A review of the literature. *Psychiatric Services, 53*, 1390-1396.

Becker, G., & Kaufman, S. (1988). Old age, rehabilitation, and research: A review of the issues. *Gerontologist, 28*(4), 459-468.

Bergman-Evans, B. (2004). Nursing homes, not nursing hospitals. *Journal of Gerontological Nursing, 30*(6), 3.

Blazer, D. (2003). Depression in late life: Review and commentary. *The Journals of Gerontology: Series A, Biological Sciences and Medical Sciences, 58*, 249–265.

Bowlby-Sifton, C. (2000). Maximizing the functional abilities of persons with Alzheimer's disease and related dementias. In M. P. Lawton & R. L. Rubenstein (Eds.), *Interventions in dementia care: Toward improving quality of life* (pp. 11-37). New York: Springer.

Boyle, V. L., Roychoudhury, C., Beniak, R., Cohn, L., Bayer, A., & Katz, I. (2004). Recognition and management of depression in skilled-nursing and long-term care settings: Evolving targets for quality improvement. *American Journal of Geriatric Psychiatry, 12*(3),288-295.

Buchanan, R. J., Choi, M., Wang, S., & Ju, H. (2004). End-of-life care in nursing homes: Residents in hospice compared to other end-stage residents. *Journal of Palliative Medicine, 7*(2), 221-232.

Burrows, A. B., Satlin, A., Salzman, C., Nobel, K., & Lipsitz, L. A. (1995). Depression in a long-term care facility: Clinical features and discordance between nursing assessment and patient interviews. *Journal of the American Geriatrics Society, 43*(10), 1118-1122.

Camp, C. J. (2000). Clinical research in long term care: What the future holds. In V. Molinari (Ed.), *Professional psychology in long term care: A comprehensive guide* (pp. 401-423). New York, NY: Hatherleigh.

Carp, D. R., & Carp, H. J. (1984). Denial and other emotional responses to infertility. *Midwife Health Visit Community Nurse, 20*(8), 286-289.

Casciani, J. M. (2003). Advancing the physical well-being of older adults: Mental health and primary care. *Vericare Monograph 2*, San Diego, CA.

Centers for Medicare and Medicaid Services. (2004). *MDS active resident information report: Fourth quarter 2004*. Retrieved March 8, 2005 from http://www3.cms.hhs.gov/states/mdsreports/res2.asp.

Chambless, D. L., & Ollendick, T. H. (2001). Empirically supported psychological interventions: Controversies and evidence. *Annual Review of Psychology, 52*, 685-716.

Clegg, A., Bryant, J., Nicholson, T., McIntyre, L., De Broe, S., Gerard, K., et al. (2002). Clinical and cost-effectiveness of donepezil, rivastigmine and galantamine for Alzheimer's disease. *International Journal of Technology Assessment in Health Care, 18*, 497-507

Cody, M., Beck, C., & Svarstad, B. (2002). Mental health services in nursing homes: Challenges to the use of nonpharmacologic interventions in nursing homes. *Psychiatric Services, 53*, 1402-1406.

Cohen, C. I., Hyland, K., & Kimhy, D. (2003). The utility of mandatory depression screening of dementia patients in nursing homes. *The American Journal of Psychiatry, 160*, 2012-2017.

Cohen-Mansfield, J. (2001). Nonpharmacologic interventions for inappropriate behaviors in dementia: A review, summary, and critique. *American Journal of Geriatric Psychiatry, 9*(4), 361-381.

Courtney, C., Farrell, D., Gray, R., Hills, R, Lynch, L., Sellwood, E., et al. (2004). Long-term donepezil treatment in 565 patients with Alzheimer's disease (AD2000): Randomised double-blind trial. *Lancet, 363*(9427), 2105-2115.

Datto, C. J., Oslin, D. W., Streim, J. E., Scheinthal, S. M., DiFilippo, D., Katz, I. R. (2002). Pharmacologic treatment of depression in nursing home residents: A mental health services perspective. *Journal of Geriatric Psychiatry and Neurology, 15*(3), 141-146.

Dening, T., & Bains, J. (2004). Mental health services for residents of care homes. *Age and Aging, 33*, 1-2.

Erkinjuntii, T., Kurz, A., Gauthier, S., Bullock, R., Lilienfield, S., Damaraju, C. (2002). Efficacy of galantamine in probable vascular dementia and Alzheimer's disease combined with vascular disease: A randomized trial. *Lancet, 359*, 1289–1290.

FDA Alert (2005): Increased Mortality in Patients with Dementia-Related Psychosis. http://www.fda.gov/cder/drug/InfoSheets/HCP/quetiapineHCP.htm.

Feil, N. (1999). Current concepts and techniques in validation therapy. In M. Duffy (Ed.), *Handbook of counseling and psychotherapy with older adults* (pp. 590-613). New York, NY: Wiley.

Feldt, K. S. (2004). The complexity of managing pain for frail elders. *Journal of the American Geriatrics Society, 52*(5), 840-841.

Ferrell, B. A., Ferrell, B. R., & Rivera, L. (1995). Pain in cognitively impaired nursing home patients. *Journal of Pain Symptom Management, 10*(8), 591-598.

Foa, E. B., & Meadows, E. A. (1997). Psychosocial treatments for posttraumatic stress disorder: A critical review. *Annual Review of Psychology, 48*, 449-480.

Freis, J., Koop, C., Beadle, C., Cooper, P., England, M., Greaves, R., et al. (1993). Reducing health care costs by reducing the need and demand for medical services. *New England Journal of Medicine, 329*, 321-325.

Grossberg, G. T., & Desai, A. K. (2003). Management of Alzheimer's Disease. *The Journals of Gerontology: Series A: Biological Sciences and Medical Sciences, 58*(4), M331–353.

Hall, G. R., & Buckwalter, K. C. (1987). Progressively lowered stress threshold: a conceptual model for care of adults with Alzheimer's disease. *Archives of Psychiatric Nursing, 1*(6), 399-406.

Heath, J., Gartenberg, M., & Beagin, E. (2006). Blending mental health services into geriatric medical care of long-term care residents. In L. Hyer & R. Intrieri (Eds.), *Geropsychological interventions in long-term care* (65–84). New York: Springer.

Hyer, L. (2004). *Integration of psychiatric and psychological care in LTC*. Presentation at the Annual Meeting of the American Psychological Association. Honolulu, Hawaii.

Hyer, L., Carpenter, B., Bischmann, D., & Wu, H. (2005). Assessment and Treatment of depression in long term care. *Clinical Psychology: Science and Practice, 12*(3), 280-299.

Hyer, L. A., & Ragan, A. M. (2002). Training in long-term care facilities: Critical issues. *Clinical Gerontologist, 25*(3-4), 197-237.

Hyer, L., Aisen, P., & Rehok, G. (2006). Augmentation of standard pharmacotheraphy for Alzheimer's Disease with cognitive rehabilitation. NIA, ROI.

Kahana, B., & Kahana, E. (1982). Clinical issues of middle age and later life. *Annals of the American Academy of Political and Social Science, 464*, 140-161.

Kasckow, J. W., Mulchahey, J. J., & Mohamed, S. (2004). The use of novel antipsychotics in the older patient with neurodegenerative disorders in the long-term care setting. *Journal of American Medical Directors Association*, July–August, 242–248.

Katz, I. R., Jeste, D. V., Mintzer, J. E., Clyde, C., Napolitano, J., Brecher, M. (1999). Risperidone study group. Comparison of risperidone and placebo for psychosis and behavioral disturbances associated with dementia: a randomized, double-blind trial. *Journal of Clinical Psychiatry, 60*, 107-115.

Kaufer, D., Borson, S., & Sadik, M. (2005, March). *Alzheimer's disease: Response to cholinesterase inhibitors*. Paper presented at the American Association of Geriatric Psychiatry, San Diego, CA.

Kazdin, A. E. (1994). Methodology, design, and evaluation in psychotherapy research. In A. E. Bergin & S. L. Garfield (Eds.), *Handbook of psychotherapy and behavior change* (4th ed., pp. 19-71). New York: Wiley.

Kitwood, T. (1995). Exploring the ethics of dementia research: A response to Berghmans and Ter Meulen: A psychosocial perspective. *International Journal of Geriatric Psychiatry, 10*(8), 655-657.

Kitwood, T., & Benson, S. (1995). *The new culture of dementia care.* London: Hawker Publications.

Kovach, C., Taneli, Y., Dohearty, P., Schlidt, M., Cashin, S., & Silva-Smith, A. (2004). Effect of the BACE intervention on agitation of people with dementia. *The Gerontologist, 44*(6), 797-806.

Lawton, M. P. (1999, 23 August). *Measuring quality of life in nursing homes: The search continues.* Invited address, divisions 34, 5, & 20, Annual meeting of the American Psychological Association, Boston, MA.

McGilton, K. S. (2002). Enhancing relationships between care providers and residents in long-term care: Designing a model of care. *Journal of Gerontological Nursing, 28*(12), 13-21.

McKeith, I., Del Ser, T., & Spano, P., Enre, M., Wesness, K., Anand, R., et al. (2000). Eficacy of rivistigmine in dementia with Lewy bodies: A randomized, double blind, placebo-controlled interventional study. *Lancet, 356,* 2031-2036.

Miller, S. C., Mor, V., Teno, J. (2003). Hospice enrollment and pain assessment and management in nursing homes. *Journal of Pain & Symptom Management, 26*(3), 791-799.

Mintzer, J., Colenda, C., Waid, L., Lewis, L., Meeks, A., Stuckey, M., Bachman, D., Saladin, M., & Sampson, R. (1997). Effectiveness of a continuum of cafre using brief and partial hospitalization for agitated dementia patients. *Psychiatric Services, 48,* 2031–2036.

Mitnitski, A. B., Song, X., Rockwood, K. (2004). The estimation of relative fitness and frailty in community-dwelling older adults using self-report data. *Journal of Gerontology Series A Biological Sciences and Medical Sciences,* 59, M627-M632.

Mold, J. W., Fryer, G. E., & Roberts, A. M. (2004). Why do older patients change primary care physicians? *Journal of the American Board of Family Practice, 17*(6), 453-460.

Murphy, C. F., & Alexopoulos, G. S. (2004). Longitudinal association of initiation/perseveration and severity of geriatric depression. *The American Journal of Geriatric Psychiatry,* 12, 50-56.

Nanna, M. J., Lichtenberg, P. A., Buda-Abela, M., & Barth, J. T. (1997). The role of cognition and depression in predicting functional outcome in geriatric medical rehabilitation patients. *Journal of Applied Gerontology,* 16, 120-132.

Olin, J., & Schneider, L. (2002). Galantamine for Alzheimer's Disease. *Cochran Database System Review,* CD001747.

Omnibus Budget Reconciliation Act (OBRA) of 1987, Public Law No. 100–203, Title IV, subtitle C, sections 4201–4206, 4211–4216, 101 Stat 1330–160 through 1330–220, 42 USC section 1395; -3(a)–(h) [Medicaid] (1992).

Ormel, J., Van cen Brink, W., & Koeter, M. (1995). Recognition, management, and outcome, of psychological disorders in primary care: A naturalistic follow-up study. Chichester, England: Wiley.

Parmelee, P. A., Katz, I. R., & Lawton, M. P. (1991). The relation of pain to depression among institutionalized aged. *Journals of Gerontology,* 46, 15-21.

Parmelee, P. A., Katz, I. A., & Lawton, M. P. (1992). Incidence of depression in long-term care settings. *Journals of Gerontology,* 47, M189-M196.

Peskin, E. (2003). Management of depression in long-term care of patients with Alzheimer's disease. Journal of American Medical Directors Association (Nov–Dec), S141–S145.

Porock, D., Oliver, D., Zweig, S., Rantz, M., Mehr, D., Madsen, R., et al. (2005). Predicting death in nursing homes: Development and validation of the 8-month Minimum Data Set Mortality Risk Index. *Journal of Gerontology, 60A*(4), 491-498.

Porteinsson, A. P., Tariot, P., & Erb, R. (2001). Placebo-controlled study of divalproex sodium for agitation in dementia. *American Journal of Geriatric Psychiatry,* 9, 58-66.

Radloff, L. S. (1977). The CES-D scale: A self-report depression scale for research in the general population. *Journal of Applied Psychosocial measures,* 1, 385-401.

Reisberg, B., Doody, R., Stoffler, A., Schmitt, F., Ferris, S., & Mobius, H. (2003). Memantine in moderate to severe Alzheimer's disease. *New England Journal of Medicine,* 348, 1333-1341.

Rhoades, J. A., & Krauss, N. A. (1999). *Nursing home trends, 1987 and 1996 MEPS Chartbook No. 3.* (AHCPR Pub. No. 99-0032). Rockville, MD: Agency for Health Care Policy and Research, Public Health Service, US Department of Health and Human Services.

Rovner, B. W., German, P. S., Broadhead, J., Morriss, R. K., Brant, L. J., Blaustein, J., et al. (1990). The prevalence and management of dementia and other psychiatric disorders in nursing homes. *International Psychogeriatrics*, 2(1), 13-24.

Ruckdeschel, K., Thompson, R., Datto, C. J., Streim, J. E., & Katz, I. R. (2004). Using the Minimum Data Set 2.0 mood disturbance items as a self-report screening instrument for depression in nursing home residents. *The American Journal of Geriatric Psychiatry*, 12, 43-50.

Ryan, M. J., Kidder, S., Daiello, L., & Tariot, P. (2002). Mental health services in nursing homes: Psychopharmacologic interventions in nursing homes: What do we know and where should we go? *Psychiatric Services*, 53, 1407-1413.

Sakauye, K. M., Camp, C. J., & Ford, P. A. (1993). Effects of buspirone on agitation associated with dementia. *American Journal of Geriatric Psychiatry*, 1, 82-84.

Schneider, L. S., Dagerman, K., & Insel, P. S. (2006). Efficacy and adverse effects of atypical antipsychotics for domentia: meta-analysis of randomized, placebo-controlled trials. Am J Geriatric Psychiatry, 14(3):191-210.

Schneider, L. S., Tariot, P. N., Dagerman, K. S., Davis, S. M., Hsiao, J.K., Ismail, M.S., et al. (2006). Effectiveness of atypical antipsychotic drugs in patients with Alzheimer's disease. *New England Journal of Medicine*, 355(15), 1525-1538.

Schulz, R., Belle, S., Czaja, S. J., McGinnis, K. A., Stevens, A., Zhang, S. (2004). Long-term care placement of dementia patients and caregiver health and well-being. *Journal of the American Medical Association*, 292(8), 961-967.

Shelton, P. S., & Hocking, L. B. (1997). Zolpidem for dementia-related insomnia and night time wandering. *The Annals of Pharmacotherapy*, 31(3), 319-322.

Silver, J. M., Kahn, D. A., Frances, A., Carpenter, D. (1998). The expert consensus guidelines™: Agitation in older persons with domentia. A postgraduate medicine special report April 1998; www.psychguides.com/ecgsl.php.

Sink, K., Holden, K., & Yaffe, K. (2005). Phramocological treatment of neuropsychiatric symptoms of dementia. *Journal of the American Medical Association*, 293, 596-608.

Snow, L., Powers, D., & Liles, D. (2006). Cognitive-behavioral therapy for long-term care patients with dementia. In L. Hyer & R.Intrieri (Eds.), *Geropsychological interventions in long-term care (265–294)*. New York: Springer

Stefanacci, R. G. (2005). Assisted living consult: What's in a name? Assisted Living Consult 2005, 1(1), 6-7.

Strain, J., Lyons, J., & Hammer, J. (1991). Cost-offset from a psychiatric consultation- liason intervention with elderly hip fracture patients. American Journal of Psychiatry, 148, 1044-1049.

Sutor, B., Rummans, T., & Glenn, S. (2001). Assessment and management of behavioral disturbances in nursing home patients with dementia. *Mayo Clinic Proceedings*, 76(5), 540-550.

Tariot, P. N., Cummings, J. L., Katz, I. R., Mintzer, J., Perdomo, C., Schwam, E., et al. (2001). A randomized, double-blind, placebo-controlled study of the efficacy and safety of donepezil in patients with Alzheimer's disease in the nursing home setting. *Journal of American Geriatrics Society*, 49(12), 1590-1599.

Tariot, P., Farlow, M., Grossberg, G., Graham, S., McDonald, S., & Gergel, I. (2004). Memantine treatment in patients with moderate to severe Alzheimer's disease already receiving donepezil: A randomized controlled trial. *Journal of the American Medical Association, 291*, 317-324.

Tariot, P. N., Rosemary, R. N., Podgorski, C. A., Cox, C., Patel, S., Jakimovich, L., et al. (1998). Efficacy and tolerability of carbamazepine for agitation and aggression in dementia. *The American Journal of Psychiatry, 155*(1), 54-61.

Teri, L., Gibbons, L. E., McCurry, S. M., Logsdon, R. G., Buchner, D. M., Barlow, W. E., et al. (2003). Exercise plus behavioral management in patients with Alzheimer Disease: A randomized controlled trial. *Journal of the American Medical Association, 290*(15), 2015-2022.

Teri, L., Logsdon, R. G., Peskind, E., Raskind, M., Weiner, M. F., Tractenberg, R. E., et al. (2000). Treatment of agitation in AD: A randomized, placebo-controlled clinical trial. *Neurology, 55*(9), 1271-1278.

Teri, L., McKenzie, G., & LaFazia, D. (2005). Psychosocial treatment of depression in older adults with Dementia. Clinical Psychology: *Science and Practice, 12*(3), 303-316.

Trinh, N., Hoblyn, J., Moharty, S., & Yaffe, K. (2003). Efficacy of cholinesterase inhibitors in the treatment of neuropsychiatric symptoms and functional impairment in Alzheimer's disease: A meta-analysis. *Journal of the American Medical Association, 289*, 210-216.

U.S. Census Bureau. (2002). *National long term care survey*. Retrieved December 28, 2007. http://www.census.gov/prod/1/pop/p23–190/p23190-e.pdf.

Verhaeghen, P., Marcoen, A., & Goossens, L. (1992). Improving memory performance in the aged through mnemonic training: A meta-analytic study. *Psychology & Aging, 7*(2), 242-251.

Volicer, L., & Hurley, A. C. (2003). Management of behavioral symptoms in progressive degenerative dementias. *Journals of Gerontology: Series A: Biological Sciences & Medical Sciences, 58A*(9), 837-845.

West, R. L., Welch, D. C., & Yassuda, M. S. (2000). Innovative approaches to memory training for older adults. In R. D. Hill, L. Backman, & A. Stigsdotter Neely (Eds.), *Cognitive rehabilitation in old age* (pp. 81-105). New York: Oxford University Press.

Williams, T. F. (1998). Comprehensive geriatric assessment. In E.H. Duthie, Jr. & P. R. Katz (Eds.), *Practice of geriatrics* (3rd ed., pp. 15-22). Philadelphia: WB Saunders.

Zgola, J. (1999). Ongoing appraisal of the person's cognitive abilities. In J. Zgola (Ed), *Care that works* (pp. 44-50). Baltimore: Johns Hopkins University Press.

Chapter 7
Professional Practice: Disciplines, Documentation, and Outcomes Measurement

Merla Arnold, John C. Colletti, and Nicholas C. Stilwell

Introduction

While the old and very old have been attended to by health care providers for centuries, the current climate for providing mental health services to older adults has primarily taken shape over the last 20–35 years. Since the deinstitutionalization of people with serious and persistent mental illness took place in the 1970s, the primary care specialties that have been treating larger numbers of older adults with mental health problems in the community, nursing homes, and assisted-living settings have been psychologists, social workers, nurses, and psychiatrists (and all too often, the primary care physician).

Each discipline has its own guidelines, rules, regulations and training paradigms, in addition to clinical roles and responsibilities. The first section of this chapter describes each of the mental health specialties providing services to older adults in long-term care (LTC) settings. Subsequent sections review the requirements for documenting services, record keeping guidelines and recommendations, and outcomes assessment or quality measurement.

Providers of Mental Health Services

Social Workers. Clinical social workers (CSWs) have a degree in social work (usually a master's degree) and special training in counseling. There are social workers with baccalaureate degrees that work under the supervision of a CSW in nursing homes and may be the only professional having a social work background in an assisted-living setting.

Specialties in gerontology are included in many social work programs. Gerontology concentrations are among the most often chosen fields of practice for both undergraduate and graduate social workers (Johnson, Kuder, & Wellons, 1992). As many as five training models have been identified. A paper completed by Johnson (1992) et al., describes certificate programs that can provide advanced gerontological content, aging research opportunities, and multidisciplinary practicum experiences.

E. Rosowsky et al. (eds.), *Geropsychology and Long Term Care*,
DOI: 10.1007/978-0-387-72648-9_7, © Springer Science + Business Media, LLC 2009

At the University of Kentucky, for example, there is a generalist or basic social work program with an ability to select electives and practica in aging. Graduate students who enroll in the Graduate Certificate in Gerontology program must complete 22 semester hours of generalist, social work courses; 12 hours of electives from the core of approved, multidisciplinary gerontology courses; and 20 semester hours of practica within agencies serving older persons.

While there can be great variability among programs, the direction of the training of social workers appears to be moving toward a specialist position in the treatment of older adults. Social workers most often provide psychotherapy services; however, the type of treatment and the populations they serve greatly vary. As noted earlier, many nursing homes have a master's-level social worker on staff, or as an outside consultant, as a response to the requirements in the 1987 OBRA (Omnibus Reconciliation Act, 1993) laws requiring that the psychosocial needs of residents be addressed and incorporated into the residents' care plans. However, unless the stay is not under Medicare Part A, clinical and licensed social workers are not eligible for payment for inpatient services (services in skilled nursing facilities are billed as "inpatient" services) under Part B Medicare, as are psychologists (http://www.empiremedicare.com/newypolicy/policy/L3671_final.htm).

The larger skilled nursing facilities (usually greater than 125 beds) are required to have a social services director with at least a master's degree. Social workers often function as case managers and can be the discipline responsible for the mental health portion of the Minimum Data Set (MDS) required of all nursing homes (discussed in greater detail below).

Nurses. For the origin of gerontological nursing, one must look at the 1955 Social Securities Act and the resultant development of adult homes and boarding homes. Many of these institutions were owned by registered nurses. In 1966, a favorable turn was taken when the American Nurses Association (ANA) established a division of Geriatric Nursing Practice. This division spearheaded several efforts to further establish quality of service and define scope of practice. Among its accomplishments, the Division of Geriatric Nursing Practice developed A Statement on the Scope of Gerontological Nursing Practice (ANA, 1982b); A Challenge for Change: The Role of Gerontological Nursing (ANA, 1982a); and Gerontological Nursing: The Positive Difference in Healthcare for Older Adults (ANA, 1980). These cornerstone achievements have spurred a growth in gerontological nursing education.

A description of the types of nursing specialties that exist today can be found in Miller's text (1997), *Health Care Choices for Today's Consumer*. Nurse practitioners (NPs) have the training and education that permits them to serve as physician "extenders." The NP examination can include a physical examination (including gynecological and breast examinations, for example). They also conduct medical histories, diagnose, and treat minor illnesses and injuries, order and interpret lab tests and X-rays, and counsel patients. They refer patients to physicians if more expert medical attention is required. Most states allow NPs to prescribe medications, but generally under a physician's supervision or oversight.

Nearly 50,000 NPs practice in the United States, many in such specialties as adult health, family health, gerontology, obstetrics/gynecology, community health, and mental health. NPs work under the supervision of a physician and, in most states, but not all, cannot practice independently as can psychologists and master's-level licensed social workers. NPs have been successful in gaining the privilege to work independently of physicians in more than 10 states. In these states, NPs are able to order lab tests, diagnose, prescribe medications, and treat without physician involvement in the case (Pearson, 2007).

Clinical nurse specialists (CNSs) are registered nurses with master's or doctorate degrees in specialized areas such as maternal and child health, mental health, gerontology, cancer, diabetes, and cardiac care. About 58,000 CNSs are in practice. These advanced practice nurses traditionally work in hospitals but are increasingly working in HMOs and clinics, where they often provide mental health services. A CNS can provide primary care and psychotherapy, conduct health assessments, and diagnose and treat illnesses.

NPs and CNSs can bill Medicare for psychotherapy services, with the exception of psychoanalysis and hypnosis. Psychotherapy codes that include a medical evaluation and management service are available to NPs and physicians (http://www.empiremedicare.com/newpolicy/policy/L3671_final.htm).

In assisted-living settings, a baccalaureate-level nurse typically supervises associate-level licensed practical nurses (LPNs). They often work out of the facility's "wellness office." Here they monitor vital signs, give out medicines and often liaison with the aides (or personal care managers), physician(s) and other consultants, and family members. The LPN typically provides more of the "hands on" work.

Psychiatrists. Psychiatrists are physicians who specialize in diagnosing and treating mental and emotional disorders. As medical doctors, many psychiatrists embrace the biological or medical model of disease care (for a complete discussion see Chap. 6). This usually means the prescribing of medications to treat a variety of concomitant concerns more commonly found among older adults. It should be noted that a psychiatrist may or may not be the prescriber of medications to treat psychiatric disorders. Typically, it is the primary care physician who is considered the "team leader" and who, not uncommonly, prescribes psychiatric medications with, and at times without, the input from a consulting psychiatrist. Psychiatrists can bill for medication management, medical evaluation and management, and psychotherapy services.

Psychologists. Psychologists serve many roles in the provision of mental health services to older adults in LTC settings. A major distinction between psychology and other disciplines can be found in the type and scope of training, what services are reimbursed, and for how much. Psychologists are often called to consult on a case, provide psychological testing, or to devise and implement behavioral management treatment plans. These tasks are in addition to the areas of consultation that psychologists share with other professions (e.g. psychotherapy, family interventions, training, etc.).

Psychologists are graduates of doctoral programs in psychology and are licensed to practice psychology by their respective states. Psychologists are eligible to become Medicare providers and bill for psychological services (e.g., initial diagnostic evaluations, testing, psychotherapy, and health and behavior management).

Geropsychologists

Two major conferences, commonly referred to as Older Boulder I, held in 1981 and Older Boulder II, held in 1992, identified the "what" of geropsychology training (Scogin, 2007) vis-à-vis the knowledge base, and the attitudes, and skills required of a geropsychologist (Knight & Karel, 2006). The basic professional practice behaviors were suggested in *The Guidelines for Psychological Practice with Older Adults* (APA, 2004). In 1998, and again in 2005, the American Psychological Association (APA) recognized geropsychology as a proficiency area. Though a petition to APA to recognize geropsychology as a specialty was filed on behalf of APA Divisions 20 and 12, Section II (Society of Clinical Geropsychology) supporting geropsychology's specialty status, the APA Commission for the Recognition of Specialties and Proficiencies in Professional Psychology (CRSPPP, 1995) denied the application, citing the absence of a training model in geropsychology. Subsequently, the National Conference on Training in Professional Geropsychology was convened during the summer of 2006 in Colorado Springs, Colorado, out of which came the Pikes Peak Model (PPM) of geropsychology training.

The PPM defined a basic model of training designed to develop in the learner the basic competencies required for competent geropsychology practice (Knight & Karel, 2006). The Guidelines for Psychological Practice with Older Adults is reflected throughout PPM emphasizing, as noted above, the attitudes, knowledge, and skills consistent with competent geropsychology practice. The PPM does not prescribe specific course work or number of practicum hours, but it does identify core competencies required of a competent geropsychologist. In addition, the model underscores the importance of didactic training on a level consistent with graduate-level course work, with clinical training provided by competent geropsychologists.

There are many avenues to develop geropsychology competencies, including courses offered by graduate-level programs, internships, postdoctoral fellowships, and postlicensure continuing education courses and conferences (APA Division 12, Section II, 1997). A 2006 directory of pre- and postdoctoral training opportunities in Clinical Geropsychology (available on the website of the Society of Clinical Geropsychology at http://geropsych.org), lists over 100 internship and postdoctoral fellowship sites and 13 graduate programs, offering substantive clinical training in the field. Additionally, opportunities for obtaining continuing education in this area continue to expand.

Generalist psychologists and other health care professionals can make good use of the expertise that geropsychologists (psychologists with advanced training and experience in mental health and aging) bring to cases complicated by aging-related factors, beyond that which the generalist might typically see or have trained for. In the opinion of many, the advanced practitioner in clinical geropsychology warrants credentialing from a specialty board, like those offered in other specialties by the American Board of Professional Psychology (ABPP) and similar credentialing bodies. At this time, however, an ABPP in geropsychology is not available.

Nonetheless, to work competently with older individuals, their families, and caregivers, psychologists require not only solid, general preparation in psychological sciences but also additional training specific to aging-related issues. Such an approach builds on psychology's fundamental training and assumes a firm grounding in core content areas (such as biological bases of behavior, cognitive, intrapersonal, interpersonal, and environmental factors), applied knowledge and skills (such as psychopathology, assessment, treatment, ethics), and research knowledge and skills (e.g., Teri, Storandt, Gatz, Smyer,&Stricken, 1992). Geropsychologists in LTC settings can and do provide key services as members of a comprehensive interdisciplinary team.

Documentation in the Medical Record

All mental health practitioners providing services to older adults in LTC settings should be familiar with the policies and procedures for entering necessary information into the medical record. While clinicians should maintain their own clinical records, the requirements of different individual practice settings can run the gamut from assisted-living sites preferring no documentation, to a need to document full assessments, treatment plans, ongoing signs and symptoms, response to interventions, and changes or updates to the plan. Documentation of assessment and treatment serves several purposes and is a requirement if the practitioner is billing Medicare Part B (see below). Some reasons for documentation relate to practical clinical issues such as maintaining continuity of care or direction of treatment, measuring progress, and communicating relevant and Health Insurance Portability and Accountability Act (HIPAA) compliant treatment issues to other members of the health care team.

Clinical documentation is taught as a matter of practice and therefore most psychologists entering practice in LTC should be able to document their efforts. For details, see the most recent update of the APA's Record Keeping Guidelines (2007) at <http://www.apa.org/practice/recordkeeping.pdf>. A more challenging aspect of documentation concerns not only managing practice-related communications but also understanding the requirements of a particular facility, or a third-party insurance carrier, in conjunction with adherence to state and federal rules and regulations. Adherence to these rules and regulations is necessary if the psychologist wants the privilege to practice in an LTC settings and receive reimbursement for work done within these systems (see Chap. 1).

Currently, the major source of reimbursement for work done in LTC is through Medicare Part B. Medicare is a federal health program for seniors administered by directives from the Centers for Medicare and Medicaid Services (CMS) that are carried out locally by the states through specific insurance carrier agent(s). Insurance carriers have drafted policies to satisfy CMS directives through Carrier Advisory Committees (CACs) that provide input in the development of local medical review policies (LMRPs).

CMS instructed the carriers to convert all existing LMRPs into local coverage determinations (LCDs) consisting of information related to "reasonable and necessary" criteria, diagnoses supporting reasonable and necessary standards, and documentation requirements. The LMRPs/LCDs may be more specific than CMS directives. These policies outline what clinicians must document to demonstrate that treatment is medically necessary and to receive reimbursement for their work (medical necessity will be addressed in a later section). An example LCD for Mental Health Services in New York State can be found at <http://www.empiremedicare.com/newypolicy/policy/L3671_final.htm>.

Clinicians providing services in LTC settings need to be educated about, and keep current with, these policies, rules, and regulations in order to help ensure that documentation includes the "acceptable" and required components of all evaluation and treatment efforts. The nation's LMRPs/LCDs, sorted by insurance contractors and states they serve, can be viewed at <http://www.cms.hhs.gov/Determination Process/04_LCDs.asp#TopOfPage>. Draft LCDs can also be viewed at local insurance contractor Web sites. Check these sites frequently for updates. Carrier newsletters also help practitioners stay current.

The next sections of this chapter review the basic issues regarding documentation under the Medicare system. The general approaches reviewed may also reflect the requirements of other third-party payers and regulatory systems, since many carriers follow Medicare guidelines. To verify documentation requirements, be sure to review your own carrier's rules and regulations.

The psychologist should also be aware of documentation standards that are covered under protected health information (PHI) requirements, as set forth under the HIPAA. The HIPAA "Privacy Rule" compliance date for meeting these regulation standards was April 14, 2003. The APA Practice Directorate produced a product that assists psychologists in understanding and complying with HIPAA regulations (http://www.apapractice.org/apo/hipaa.html#). The reader may also acquire additional information by referring to the Department of Health and Human Services, Office of Civil Rights (HHS/OCR) Web site (www.hhs.gov/ocr/index.html). The final HIPAA Privacy Rule, as well as a very informative list of frequently asked questions, can be viewed at the HHS/OCR site.

Psychologists falling under HIPAA rules are required to, among other things, provide a Notice of Privacy Practices to the patient, by the first date of service. Psychologists are also required to treat the guardian or legal representative (the person with legal authority to make health care decisions for the beneficiary or patient) as the patient with respect to PHI. Psychologists working under an Organized Health Care Arrangement may be included in a single Joint Notice of Privacy Practices given to the patient by the long-term care facility. It is beyond the scope of this chapter to address implementation of practices to meet HIPAA regulations. Readers should consult the resources noted, as well as their state regulations and relevant practice organizations for guidance.

Documentation in the clinical record can aid in the measurement and evaluation of clinical progress and outcomes, thereby facilitating reviews of treatment strategies, approaches, and the overall effectiveness of treatment plan modifications. Effective documentation can help clinicians target the most positive, efficient, and effective approaches and outcomes from both the residents' and clinicians' perspectives. On a larger scale, outcomes measurement provides support for the value of psychological assessment and treatment for older adults living in LTC settings.

Documenting progress toward a treatment goal can also provide evidence of treatment effectiveness and need for continuing services. Although Medicare does recognize there will be treatment impasses, in today's practice environment, lack of progress can be equated with lack of medical necessity for additional treatment. In the event of the need for ongoing maintenance therapy, the key is to document in a way that demonstrates that the withdrawal of treatment is likely to exacerbate symptoms. It is also useful to document what (if any) alternative treatment approaches are under consideration. Treatment that is necessary to prevent relapse or to maintain functioning is a reimbursable service under Medicare. The critical element is clear documentation supporting this judgment.

Again, clinicians in LTC who submit claims for reimbursement to Medicare must document the "medical necessity" of the treatment provided. As noted, under Part B Medicare, mental health services must be "reasonable and necessary for the diagnosis, or for treatment of an illness or injury, or to improve the functioning of a malformed body member" (Office of Inspector General, 1996). That is, "individual psychotherapy meets medical necessity criteria if the patient has a psychiatric illness and/or demonstrates emotional or behavioral symptoms that cause inappropriate behavior or maladaptive functioning" (HGSA Administrators, 2005).

In New York's LCD, for example, reasonable and necessary includes a "reasonable expectation of improvement…demonstrated by an improved level of functioning, or maintenance of level of functioning where decline would otherwise be expected in the case of a disabling mental illness or condition or chronic mental disorders" (http://www.empiremedicare.com/newypolicy/policy/L3671_final.htm). A person who cannot actively participate or benefit from psychotherapy does not meet the criteria for medical necessity.

If psychotherapy is provided to a patient with dementia, the patient's record should make clear that the patient's cognitive level of functioning is sufficient to permit the patient to meaningfully participate in the treatment. Documenting the ability to develop, or that the patient has developed, a therapeutic relationship can serve to demonstrate medically necessary and appropriate services, and then make reimbursement more likely when treating a person with more advanced dementia.

Relating medical necessity to outcomes measurement suggests that clinicians should be able to document progress toward or attainment of goals. Such positive outcomes include a reduction of symptoms such as inappropriate behavior, maladaptive functioning, or subjective distress. The focus is on documenting effective goal-oriented interventions with measurable outcomes. The increasing demand for specificity in documenting effectiveness of goal-oriented therapies also applies to third-party payers other than Medicare, including managed care organizations.

This chapter will later address specific documentation issues and approaches to take when measuring or assessing treatment outcomes. We hope to provide a basic understanding of documentation concepts and outcomes issues to provide a framework within which a psychologist can document the necessity and efficacy of needed mental health services for older adults in LTC settings.

Documenting Services

When thinking about documenting consultative and/or clinical services in LTC settings, the best way to think about this may be to keep in mind the processes of care namely: (1) the referral (who, why), (2) information gathered during the assessment, along with relevant clinical formulation, (3) treatment plan, including goals and objectives (steps to the goal), (4) treatment progress and review (with modified plan as needed), and (5) considerations of discharge from services, including follow-up, as needed. Considering this format, specific documentation needs can be identified that can in turn serve to guide the ongoing development of the clinician's approach to the task.

The approaches to documentation noted below in greater detail are oriented toward record keeping in an inpatient, i.e., skilled nursing home environment as the requirements are most stringent in these settings. The approaches to documentation discussed consist of content from CMS policies, a local carrier's recommendations (HGSA Administrators, 2005) and the experiences of psychologists practicing in LTC.

Referral. In a skilled nursing home setting, prior to evaluating a resident, an order on the medical record is required from the resident's attending physician or NP working under a physician's supervision, or physician named by family or the resident, indicating that she or he has evaluated the resident's symptoms and authorizes the procedure. The physician's order will state the symptoms or problem to be evaluated. "As needed" or PRN orders have been overused or abused under Medicare and are no longer allowed in skilled nursing facilities. Information on the referral process is addressed in an Chap. 2.

It is necessary for the psychologist to not only document assessment and treatment efforts but to also be aware of sources of information and types of documentation that can demonstrate the clinical service is necessary and appropriate. One source of information available to the psychologist to help determine the "medical necessity" for the service is the facility's chart, for example. If the resident's chart does not demonstrate clear documentation of the emotional/behavioral/cognitive symptoms that necessitate clinical intervention, the medical necessity of the service could be questioned. In these circumstances, it is particularly important that the clinician's notes provide that documentation. Often, needed corroboration can be found in notes by dietary (appetite level), physical therapy (motivation level), nursing (somatic and other "complaints"), among others.

Clinical Evaluation. In the Medicare system, and most settings functioning under the medical model, the initial clinical evaluation is referred to as the diagnostic interview. The diagnostic interview often includes documentation of medical and psychiatric history, a complete mental status examination, a diagnosis of the problem, and an evaluation of the patient's ability and motivation to address the problem in therapy.

The diagnostic interview may include several or all of the following specific information, depending on the requirements of local insurance carriers: (1) the reason for the clinical interview; as noted previously, the reason for the referral should be consistent with what is documented in the chart in order to help support medical necessity; (2) a referral source and history of the problem; (3) history – medical, psychiatric, social, and family; (4) current medications; (5) clinical observations and a mental status examination; (6) functional strengths and deficits; (7) diagnostic impression or multiaxis diagnosis coded to the highest level of specificity (using the ICD-9 coding system and not the DSM IV-TR); and (8) treatment plan and recommendations. When providing services in LTC settings, it is important to obtain information from several sources, including family, staff, and other team members, in addition to utilizing the facility's medical record. New York allows that, "in certain circumstances other informants may be seen in lieu of the patient." It is always important to check the local LCD (http://www.cms. hhs.gov/DeterminationProcess/04_LCDs.asp#TopOfPage).

In sum, documentation of the diagnostic interview should convey that there was a review of the record, communication with staff and/or other informants, evaluation of the resident, and a diagnostic assessment with a diagnostic formulation or impression. From this formulation, treatment plans or intervention recommendations are developed and communicated to the attending physician and other team members. Documentation in the record is generally considered sufficient evidence of this communication. It is recommended that clinicians follow the most stringent documentation rules or requirements, as this will increase the likelihood of a positive outcome should there be a third-party audit of the clinician's record.

Psychological and Neuropsychological Testing. The Office of Inspector General's (OIG) report on mental health services in nursing homes (1996) identified testing as an area where medically unnecessary services were being provided. The 2001 follow-up report noted that psychological testing continued to be the most problematic nursing home service reviewed (OIG, 2001). Concern about inappropriate use of testing services underscores the importance of clear documentation of need. In general, it will be important to document the indications for the diagnostic tests, tailored to the individual's specific suspected problem, and that they are necessary for the diagnosis and/or the treatment plan (www.empiremedicare.com/newypolicy/policy/L3671_final.htm#29).

The medical necessity of each test may require documentation, as standardized batteries may not meet medically necessary standards. Additional pertinent documentation includes the referral information, including the reason for the referral, initial evaluation by the psychologist determining the need for a specific type of testing, results of previous testing, if any; the total testing time, tests administered, diagnosis, and recommendations resulting from the testing results. Again, it is strongly recommended that psychologists check with their local carriers and their local carrier determinations to help ensure that documentation requirements are met.

Psychotherapy Notes. The general considerations applicable when writing psychotherapy progress notes also apply to LTC settings. Therapy notes are written following each session in as timely fashion as possible. The notes should be legible and written in a language understandable by the other members of the treatment team, third-party payers, and regulatory agencies.

In addition to the above general considerations, there are more specific issues that the psychologist will want to keep in mind when documenting therapy or other psychological services provided in an LTC setting. There are important considerations when the psychologist, who is no longer working individually, in a private practice office, but rather providing services within a health care facility, which may or may not be part of a health care corporation, which operates within a state system, which responds to directives from a federal system, which is responsive to the consumer community, which tries to represent the needs of the individual elder and their families.

Basically, the psychologist must be aware that she or he is documenting a therapeutic service that is meant to fit into and be consistent with the overall plan of care for the resident that is implemented by multiple disciplines and responsive to multiple regulations. For a clinical example, if the therapy notes reflect that the resident is evidencing less-frequent episodes of depression, utilizing improved problem-solving skills and reversing a previous significant weight loss, the documentation from other disciplines should not be contradictory. That is, it should not reflect, for example, increased verbal or physical aggression, decreased meal completion, and expression of a wish to die. If such contradictory information is present, it warrants a clinical explanation or hypothesis and a plan to address. The psychologist needs to be aware of the overall plan of care. In addition, it is useful to document efforts made to communicate with other disciplines and integrate psychological treatment with other disciplines' interventions (e.g., "discussed with nursing…, spoke with resident's physician regarding…"), where appropriate.

Psychotherapy notes should also address the medical necessity of the service being provided. As such, the psychologist may want to approach each note as if it is going to be read by a third party (physician, regulatory agency, Medicare carrier) and as an isolated complete record of the therapy process. That is, the note documents where the patient has been (i.e., baseline or previous functioning level), where you and the patient are in the treatment process, and where you expect the treatment process to go (expected outcomes or goals).

Documentation reflecting treatment that is both necessary and appropriate often includes the emotional and behavioral symptoms or a psychiatric illness affecting functioning; duration of the session; the symptoms addressed; the specific therapeutic interventions that were implemented; the patient's response to the interventions; an estimate of the duration of sessions required to complete the treatment; diagnosis; therapeutic goals and methods of monitoring the outcome of treatment; and plans for next session and communication with other disciplines regarding the case.

A statement regarding the patient's capacity to be a participant in and benefit from the psychotherapy may also be necessary, particularly in the presence of dementia. The Office of Inspector General states in their report *Medicare Payments for Psychiatric Services in Nursing Homes: A Follow-up* (OIG, 2001) that their review of documentation found two main keys that determine lack of medical necessity. The first relates to the provision of psychiatric services to individuals who have limited cognitive ability and therefore a presumed decreased ability to benefit from the service. The second relates to inappropriate frequency or duration of a service; a service is provided too many times, the length of the service is too long, or the course of treatment lasts more than 6 months. The psychologist may want to document the verbal and cognitive capability of that individual to "do the work" in therapy in order to move toward a treatment goal. This documentation will help address the criteria that services are necessary and there is a reasonable expectation of response to treatment.

It is also important to be aware of what not to include in a psychotherapy progress note. The service will generally not meet the criteria for medical necessity if therapy focuses on monitoring activities of daily living, recreational therapy, social interaction or the teaching of grooming skills. The psychologist should also avoid using general statements in the progress notes such as "support provided." Instead, focus on a more detailed description of how the therapeutic intervention is addressing the problems of the LTC resident.

Again, the documentation considerations noted earlier are general approaches. It is recommended that the psychologist review the local Medicare carrier's requirements for meeting documentation/medical necessity standards. Specific documentation requirements usually exist for different therapeutic formats/interventions such as family therapy, group therapy, or a neurobehavioral status examination. It is also important to keep in mind the rules under HIPAA and what constitutes private health information.

Ending Treatment. The psychologist should include documentation of "discharge planning" or plans for discontinuing treatment. When closing a case, a final summary can serve several purposes. A written report reviews and documents the extent to which the resident responded to treatment and achieved treatment goals. It also demarcates a definite end to a particular course of treatment. If the patient's emotional and/or behavioral functioning later deteriorates, an additional course of treatment may be necessary and appropriate with the need more readily identified by other team members. The discharge (from services) summary can also be useful when arranging for or coordinating with an "after-care" plan or service.

Periodic Reviews and Outcomes Measurement. This section will review general approaches and issues regarding outcomes measurement. Except for the MDS and the CMS Physicians Quality Reporting Initiative (PQRI) discussed below, specific measures will not be addressed in depth, as it is beyond the scope of the chapter. There are several excellent reviews of outcomes measures already available and they are addressed elsewhere in this text (see Chap. 3, as well as recommended readings).

In general, psychologists must regularly monitor and document progress toward treatment goals. To measure progress associated with reduction of symptoms, maladaptive behavior, subjective distress, or inappropriate functioning, it is first necessary to document the assessment of functioning and other symptoms (foci of treatment) at baseline. Bare in mind, therapeutic progress may be documented from multiple points of view, with multiple assessment methods. For example, we can measure progress of functioning from the time of initial evaluation to the last treatment session. This can be conceptualized as the usual form of outcome measurement.

We can also document progress from one session to the next. Documentation of outcome might be for example, "According to behavioral observation notes from nursing and PT, the patient had fewer episodes of tearfulness this week than last week." Progress during the session can also be documented. Such a note might read for example, "The patient evidenced a decrease in observable symptoms (or subjective complaints) of anxiety by the end of the session." Generalization of treatment progress can be documented from the treatment session to the patient's environment.

Documenting Progress in Treatment. It may be possible to document specific progress over multiple courses of treatment. The patient may be seen for an initial course of treatment with improvement, stabilization, and discharge from services. Following an initial course of treatment, there may be a recurrence of symptoms where the older adult may return to the initial baseline level. A second course of treatment may be indicated to recapture the initial gains. It is important to document this. And, when applicable, identify and document the need for attention to and improvement in other important areas of functioning.

Each of the target areas or treatment outcomes can be assessed. Such outcome measurement can demonstrate the value of a biopsychosocial approach to LTC services by virtue of documentation of successful patient outcomes. This documentation provides further evidence that older adults have the ability to benefit from the psychological interventions. In addition, movement toward the goal offers evidence that needed services are appropriate in that they address the problem(s). Improvement after reengaging in treatment can also underscore the importance of continuing psychological services to prevent relapse and to maintain optimal functioning.

Progress in treatment should be objectively documented. When therapeutic goals have been achieved, it can represent a point in the treatment process where there is necessarily a consideration of whether to continue treatment. Consideration of discontinuing treatment is also appropriate if no further movement toward treatment goals is evident, if none can be reasonably expected, and if an increase in symptoms or diminution of functioning is not reasonably expected. In the case of no quantifiable improvement in functioning (this includes resident self-statements of subjective experience), it may indicate that the patient may not reasonably be expected to benefit from continued treatment. When services no longer meet medically necessity criteria, they no longer meet criteria for reimbursement. Under these circumstances, and absent extenuating circumstances (and so documented), treatment should be evaluated for appropriate discontinuation.

As noted, documentation of phasing-out treatment is often helpful, as it allows objective measurement of what happens to the patient's functioning, when treatment is moving toward its end. For example, it may be that in discussing the end of a treatment course, the resident's symptoms increase. This may provide evidence that ongoing treatment remains necessary and appropriate. Such an event may be more likely when the LTC resident's problem is or becomes chronic. For example, a patient diagnosed as schizoaffective, depressive type may have exhausted all manner of psychopharmacological treatment and only remains stable with monthly (or more frequent) psychotherapy sessions.

The documentation in this case should reflect that stabilization of functioning is a positive outcome of treatment that, if continued, is likely to prevent more severe depression, the recurrence of psychotic symptoms, and can minimize the need for additional or more intensive interventions, e.g., increase in or addition of medications or psychiatric hospitalization. Again, documenting the expectation of stabilization or maintenance of health status and/or functioning can serve to support the medical necessity and appropriateness of continuing the treatment. Treatment progress should be reviewed and documented at least quarterly.

Treatment outcomes can be measured within the treatment paradigm that the psychologist applies to the LTC resident's problems. Documentation should indicate movement toward the treatment goals. Such improvement evidence is often inherent in and relative to the specific paradigm or approach. For example, in a cognitive behavioral model, improvement may be measured by an increase in use of externally based problem-solving strategies vs. episodes of emotional expression in order to solve observed conflicts. However, documentation of progress should always be related to reduction of the patient's symptoms/maladaptive behavior and improvement in functioning and/or subjective experience.

Outcomes Measured in Multiple Domains. Within a particular therapeutic paradigm, specific outcomes can be measured in multiple domains. In a standard fashion, we often conceptualize outcome measurement by assessing progress in affective, cognitive, and/or behavioral domains, e.g., the person is evidencing improved mood, is less confused, and is spending increased time out of the room (or apartment). Taking a macro perspective, it may be helpful to consider what types of resources can be utilized in assessing therapy effectiveness in LTC settings.

In a 1996 article, Hans Strupp discussed the framework of the tripartite model of mental health and therapeutic outcomes, as it relates to the Consumer Reports Study on therapy effectiveness. He reviewed three perspectives for evaluating therapeutic outcomes from the perspectives of adaptive behavior (society), a sense of well-being (the older adult), and personality structure (the therapist). Each of these perspectives can be considered when conceptualizing outcome assessment/quality measurement in LTC settings.

The best measurement of progress in therapy attempts to integrate the three concepts noted earlier. Improvements in adaptive behavior can be documented and confirmed by others in the LTC environment and this includes nurse's and others' chart notes, routine behavioral tracking records that can be completed by social services, certified nursing assistants, family reports, reports of other residents, completion of standardized behavioral monitoring forms, the MDS, and physician assessments, etc.

The second perspective concerns the individual patient's evaluation of a sense of improvement in his or her situation, functioning, or experience. Outcomes from the older adult's perspective can be measured with self-report scales for depression; for example, patient's reports of improved quality of life, statements of increased resolution of adjustment issues, reports of more satisfying social relations, and statements of use of strategies discussed in therapy. The concept of medical necessity often relies heavily on these reports of increasingly adaptive behavior.

The third perspective relates to the therapist's clinical judgment about the older adult's functioning. The therapist's perspective is often used to structure the treatment process while moving toward previously, mutually established goals. Although important in documenting movement toward positive outcomes, increased focus on the therapist's perspective can be a challenge in LTC outcome evaluation.

For example, the therapist may note (and even measure) that the patient is moving from the use of internal, emotion-based coping strategies to more external, problem-based coping strategies. Such a positive movement is an outcome that can be translated into medical necessity language simply by associating the change in thinking to a reduction in emotional/behavioral symptoms or problems that had previously been of sufficient intensity or concern to cause problematic behavior, maladaptive functioning, or subjective distress (i.e., reason for treatment).

In sum, therapeutic outcomes in LTC settings can be measured by various instruments or strategies, in different domains (affective, cognitive, and/or behavioral) and from different perspectives. Documented progress toward improvement can include such indices as increased appetite, decreased psychotropic medication usage, needed reduction in elevated blood pressure, and other indicators that may be directly or indirectly affected by therapeutic efforts.

Outcomes and Professional Accountability

Beyond the measurement of a patient's progress in psychotherapy, while itself is an important variable in demonstrating the effectiveness of psychological treatment, we must also consider the question of accountability of these professional, psychological services, within the larger context of the health care system. There is a growing body of literature demonstrating not only the benefits of mental health interventions but also the decreases in other health care utilization, as an outcome of the course of psychological treatment. Significant effects have been demonstrated among medical patients with depression, decreases in hospitalization rates, and reductions in needed medical interventions with patients that have chronic pain (Hunsley, 2003).

Many constituencies, including third-party payers, the recipients of psychological services, and other health care professionals, have a keen interest in the cost effectiveness of psychological services. The more mental health practitioners document and demonstrate the favorable outcomes associated with psychological care and services, as it affects patient improvement and lower utilization rates and costs, the more integrated we will become in the overall health care system (Cummings, O'Donohue & Ferguson, 2003).

The Minimum Data Set

In November 2002, the CMS, an agency of the U.S. Department of Health and Human Services, began a national Nursing Home Quality Initiative (NHQI). The nursing home quality measures derive from resident assessment data routinely collected by nursing home staff. The data collection tool is known as the MDS. The MDS collects data on the resident's physical and clinical conditions and abilities, as well as preferences and life care wishes (http://www.cms.hhs.gov/NursingHomeQualityInits/).

An evaluation of the draft version of MDS 3.0 began in 2006, with on-site reviews of a national sample of community nursing homes, in eight states (New Jersey, Pennsylvania, Georgia, North Carolina, Colorado, Illinois, California and Texas), representing various areas of the country and with a separate sample of 20 veteran administration nursing homes. The goals for the MDS 3.0 revision was to make it more clinically relevant, while still achieving federal payment mandates and quality initiatives. Another outcome goal was to integrate selected standard scales. The final draft was published January 15, 2008 (http://www.cms.hhs.gov/ NursingHomeQualityInits/25_NHQIMDS30.asp).

As noted, the MDS is a data collection tool used to gather information across domains and disciplines of care. It is submitted and signed by an RN assessment coordinator and others. There is an evaluation of speech, hearing, and vision, with a question asking specifically about the resident's ability to communicate with others.

Section C of the MDS examines cognitive patterns using a brief mental status interview (e.g., three-word repetition, delayed recall, temporal orientation). An important consideration in nursing homes and an area of data collection on the MDS includes the assessment of the extent to which the resident's cognitive skills interfere with daily decision making. Ability scores range from independent, modified independence (with some difficulty in new situations only), moderately impaired (decisions are poor and cues/supervision is needed), to severely impaired (resident rarely or never makes decisions). There is one question regarding the ability to remember the "long past," with additional questions focusing on more recent memory such as the location of one's own room, staff names and faces, and is oriented to place. The section goes on to elicit information about the signs and symptoms of delirium and the incidence of acute mental status changes.

The resident's mood is a focus in Section D and collects data related to signs of depression or anxiety. Section E asks about behavior, both physical and verbal, and whether the behavior is self or other directed. Here, we find a question about whether the resident's behavior is considered a present risk to the resident or to others. Importantly, there is a question about whether the resident rejects care such that it impacts the resident's ability to achieve treatment goals and goals to maintain optimal health and well-being. Section F examines patient preferences, as reported by the resident or personal representative with respect to self interest and ability to manage activities of daily living and engage in meaningful activities. Functional and physical abilities are also documented on the MDS. Section G looks at bladder and bowel control, an important area for residents' quality of life and needed level of staff interventions. Incontinence is a common trigger for nursing home placement.

The MDS also documents the diseases suffered by the nursing home resident. It may seem that the mental health professional can skip this area but, as noted throughout this text, it is important for psychologists and other mental health professionals working with older adults to understand the disease states suffered by the resident and their consequences to quality of life and mental health status and visa versa. The focus includes active diagnoses such as cancer, musculoskeletal difficulties (e.g., arthritis), heart or circulation problems, neurological diagnoses (e.g., Alzheimer's disease, aphasia, cerebral palsy, CVA/TIA/stroke, dementia and others), psychiatric or mood disorder (e.g., depression, anxiety disorder, bipolar disease, schizophrenia) pulmonary disorders, and others. That is, the MDS collects data from a biopsychosocial perspective.

A full section focuses on pain management, another often suffered problem of older adults living in LTC settings, particularly those in nursing homes. Other important health conditions impacting subjective experience include respiratory difficulties such as shortness of breath, recent tobacco use, and the experience of recent episodes of nausea and vomiting. Falls are another key area for measurement, as are swallowing and nutritional status, and oral or dental status. With limited mobility among many nursing home residents, skin breakdown is a serious concern. The MDS collects data on pressure ulcer frequency and severity, in addition to outcomes of previously reported pressure ulcers. Skin ulcers are a serious risk for older adults with diabetes, heart disease, obesity, and among the frail.

Section N focuses on medications. Important key medication classes for review include antipsychotics, antianxiety medications, antidepressants, hypnotics, and anticoagulants. This focus reflects the serious risks older adults face when receiving one or more of these medications. Section O, Special Treatments and Procedures, counts other therapies providing services and including speech, occupational, physical, respiratory, recreational, and psychological (as defined as provided by "any licensed mental health professional") therapies. One of the early concerns related to nursing home care was over the use of physical restraint. The MDS tracks rates of restraints' use, to help prevent their overuse.

CMS states that the nursing home quality measures, utilizing MDS data, "are not benchmarks, thresholds, guidelines, or standards of care. [But,] are based on care provided to the population of residents in a facility, not to any individual resident [and therefore] are not appropriate for use in litigation action" (http://www.cms. hhs.gov/NursingHomeQualityInits/10_NHQIQualityMeasures.asp#TopOfPage).

The quality measures document (1) the percentage of residents physically restrained, (2) had increasing anxiety or depressed mood (looking back 7- and 30-day timeframes) and (3) the percent of residents who spent most of their time in bed or in a chair. This last measure reflects the importance of physical and social activity. The outcomes measured by these quality indicators can be positively affected by the provision of quality psychological/mental/behavioral health intervention and reflects the value of an interdisciplinary biopsychosocial approach.

Physicians Quality Reporting Initiative

In December 2006, the CMS, under the authority of the Tax Relief and Health Care Act of 2006 (TRHCA), Section 101, Title I, developed a quality reporting system known as the PQRI (http://www.cms.hhs.gov/PQRI). Eligible professionals receive a financial incentive to voluntarily participate in this quality reporting program. Eligible professionals include physicians, physician assistants, NPs, CNSs, CSWs and clinical psychologists, physical and occupational therapists, and others.

The 2008 and draft 2009 quality measures were developed under the auspices of Quality Insights of Pennsylvania (QIP), the entity contracted by CMS to implement the program. Briefly, quality screens are applied and data are collected and reported on the claim forms submitted to Medicare for services provided during a specified time frame. At this time, there are over 100 measures and psychologists and other mental health professionals can report on a subset of these, and are described in greater detail below. The incentive to participate or, "bonus payment" is subject to a cap of 1.5% of the total allowed charges billed for covered services during the specified period. The stated goal of "pay for performance [is] to encourage the provision of high-quality cost-effective care" (http://www.cms.hhs.gov/PQRI).

For each eligible patient seen during the specified reporting period, each measure has a reporting frequency requirement, e.g., report one-time only, once for each identified procedure code, once for each acute episode, per each eligible patient. The specific quality-data codes in each measure represent the clinical actions being measured and again, are captured during the administrative claims process.

Satisfactory reporting and eligibility to receive the incentive payment requires that participants report on a minimum number of measures during the reporting period. For example, psychologists must report on at least 80% of the cases in which each measure is reportable and report on a minimum of three of the mental health quality measures. To participate, clinicians must have a national provider identifier (NPI), which must be included on the claim form along with the Healthcare Common Procedure Coding System (HCPCS) codes for services or procedures provided, along with the appropriate ICD-9 CM diagnosis and quality-data codes for each measure.

Practitioners are encouraged to report on as many measures as possible to increase the likelihood of achieving successful or satisfactory reporting requirements and thus become eligible for the incentive payment. For additional information, go to the CMS PQRI Web site at <http://www.cms.hhs.gov/PQRI>.

Data collection work sheets for Phase I measures can be found at <http://www.usqualitymeasures.org/qm/measures_final/default.aspx>. The clinical information required in a particular measure is entered in the appropriate box, as are the appropriate billing and diagnostic and procedure codes. The American Medical Association PQRI-related site (http://www.ama-assn.org/ama/pub/category/17432.html) provides links to coding specification sheets that include a complete list of ICD-9 (International Classification of Diseases, ninth revision) and CPT® (Current Procedural Terminology) codes used to identify patients eligible for the measure.

Quality measurement is an interdisciplinary function with input required of physicians, nurses, psychologists, and social workers, among others. The American Psychological Association (APA) and the American Medical Association each worked in collaboration with QIP who in turn consulted with the National Committee for Quality Assurance (NCQA), CMS, and others to develop the mental health quality measures. QIP convened "expert panels" (e.g., psychologists) that worked with QIP representatives, including nurses, social workers, and physicians to develop the quality measures.

As described on the PQRI Web site http://www.usqualitymeasures.org/qm/, these technical panels included people with expertise in the issues under study and with quality measurement. The panel identified quality goals, reviewed the literature, and defined and developed measure specifications. Before available for use, QIP conducts alpha and beta testing of the measures. The process rounds out with soliciting, reviewing, and incorporating public comments to the draft quality measures. Public comment is a valuable step in the measure development process as measures have been refined and improved as a result of public comment. To meet the program's objective, a quality measure must examine an area of practice that goes beyond the minimum of what would be expected for payment of services.

The quality measures address various aspects of care such as prevention, management of chronic conditions, acute episodes of care, procedure or intervention-related care, resource utilization, and coordination of care. The initial 2008 measures specific to mental health include depression and suicidal risk, dementia screening, and codevelopment of the treatment plan. The first phase of the PQRI project included measures for both physicians and nonphysicians, but according to QIP, lacked sufficient measures that address the work of physical therapists, occupational therapists, psychologists, CSWs, and chiropractors (press release, February 8, 2008). Phase II seeks to redress this somewhat.

A new draft set of mental health quality measures for use in 2009 opened for a 30-day public comment period on February 8, 2008, and included physical health status screen, treatment plan of care for the person with cognitive impairment, complicated grief assessment in a coexisting disorder, initial assessment for posttraumatic stress disorder in a coexisting disorder, assessment for obsessive compulsive disorder in a coexisting disorder, and assessment of suicide risk in a coexisting disorder (http://www.usqualitymeasures.org/qm/measures/default.aspx).

Chapter 3 underscores the importance of a thorough assessment. The PQRI program offers another important perspective to consider during the assessment and treatment process. When performing an initial diagnostic interview, for example, it would be of use to know the quality measures' data points and develop an interview tool and/or method that captures that information.

Treatment Planning. A Phase I quality measure, patient co-development of treatment plan/plan of care, examines whether Medicare beneficiaries actively participated in the development of the treatment plan or plan of care with their health care provider. Necessary documentation includes the practitioner's signature and either the cosignature of the patient or documentation that verbal agreement was obtained from the patient or, when necessary, the authorized representative.

The measure is to be reported at least once for each unique episode of care for patients seen during the reporting period. A unique episode of care is defined by the care given for each unique diagnosis (primary ICD-9 code) during the reporting period. Valid primary ICD-9 code (e.g., diagnoses appropriate for psychological and or mental health practitioners' use) billed with the corresponding CPT service codes. The codes help identify patients who are eligible for inclusion (http://www.usqualitymeasures.org/qm/measures_final/default.aspx).

Screening for Clinical Depression. This quality measure looks at the percentage of eligible patients screened for clinical depression using a standardized tool. Examples of such depression screening tools specifically identified include the

Patient Health Questionnaire (PHQ9), Beck Depression Inventory (BDI or BDI-II), Center for Epidemiologic Studies Depression Scale (CES-D), Depression Scale (DEPS), Duke Anxiety–Depression Scale (DADS), Geriatric Depression Scale (GDS), GDS – Short Version, Hopkins Symptom Checklist (HSCL), The Zung Self-Rating Depression Scale (SDS), and Cornell Scale Screening (used when there is evidence of cognitive impairment and need for caregiver report).

In defense of the measure, the panel noted that major depression is a leading cause of disability and underscored the importance of improved recognition and diagnosis of depression among older adults. The goal of this quality measure then is to improve detection and treatment of previously underdetected incidence of depression, and therefore undertreated, so that needed treatment is offered, and thus, improving the chances for better health care outcomes (http://www.usqualitymeasures. org/qm/measures_final/default.aspx).

Physical Health Status Screen. This is a Phase II draft measure developed for 2009 implementation. The quality measure intends to identify the percentage of patients aged 65 and above that have documentation of an age-appropriate physical health status screen. The measure is designed to be reported for each initial assessment or diagnostic interview for eligible patients during the reporting period. Eligible patients include all beneficiaries aged 65 or more, with any of the diagnoses eligible for Medicare reimbursement for treatment by psychologists.

The measure underscores several important issues regarding the integral role of psychologists' attention to physical health. The rationale section of the measure notes that many biological, psychological, and social issues affecting older adults and their caregivers are not being adequately addressed (http://www.usqualitymeasures. org/qm/measures/measurelist.aspx?mid=5; note: the first author of this chapter was chair of the PQRI Phase II Psychology Expert Work Group, which developed the psychology-related 2009 quality measures).

Promoting a holistic approach to care and services (i.e., biopsychosocial) remedies current gaps in care and services; a criteria of a measure that exceeds the minimum necessary for reimbursement of services. The measure underscores that psychologists (and others) working with older adults require "distinctive knowledge in the . . . biological aspects of aging including chronic illness, terminal disease, falls . . ., nutrition, exercise, sensory changes, [and] pharmacology (e.g., pharmacodynamic changes tied to aging and problems stemming from multiple medications)" (Archival Description of Clinical Geropsychology, n.d., p. 1).

The so-called age-appropriate physical health status screen includes, but is not limited to, an assessment of sensory deficits, medical diagnoses/illness/disability, medical treatment plan adherence, substance abuse, and somatic complaints. Sensory deficits refer to impairments in hearing, vision, and taste. These are important influences in communication patterns and nutritional intake. The medical diagnosis/illness/disability data points include an assessment of health, for example deficits such as heart disease, diabetes, and arthritis, common illnesses effecting older adults. Another element of the physical health assessment includes the extent to which the senior takes medications or adheres to the medical treatment plan, as prescribed by their health care provider(s).

Problems of substance abuse can be a critical concern for older adults compromised by physical disability, an aging body, and multiple medications. For the purpose of the measure, substance abuse is defined as the use of "any legal (including prescription and over-the counter medications) or illegal substance that is causing damage to the user's physical and/or mental health and causes the person legal, social, financial or other problems including endangering their lives or the lives of others" (http://www.usqualitymeasures.org/qm/measures/measurelist.aspx?mid=5). Attention to somatic complaints such as shortness of breath, dizziness, and pain are also considered important elements of a quality physical health assessment. Incorporating a biopsychosocial approach to the initial diagnostic assessment or interview (and ongoing treatment), psychologists "can learn to distinguish between dynamic and physically (i.e., biologically, chemically) related contributors to a person's presentation and experience" (Arnold, in press).

The measure intends to reinforce the importance of an interdisciplinary, comprehensive approach to health care services. It acknowledges the important role psychology and psychologists play in the provision of comprehensive health care services to our nation's seniors. When psychologists take into consideration the impact of physical health status on presenting signs or symptoms and issues of concern, treatment or approaches can target the source of the difficulty and increase the likelihood of an effective approach, a reduction in unnecessary treatments or interventions (e.g., medications) and improved outcomes.

Cognitive Impairment. At the time of this writing, there are two quality measures related to cognitive impairment. Screening for cognitive impairments in older adults is a quality measure developed in Phase I of PQRI. Phase II built upon this process measure and developed an additional measure examining treatment planning, as the next step of the assessment and treatment process.

Screening for Cognitive Impairment. This Phase I quality measure examines the percentage of patients aged 65 and above who have documented results of a standardized cognitive impairment screen. Standardized tools are defined as those that are normed and validated for use with older adults. Such screening tools include the Clinical Dementia Rating Scale, Mini Mental Status Examination (MMSE), Global Deterioration Scale, Short Portable Mental Status Questionnaire, Clock Drawing Test, Modified MMSE, Mini-Cog, Hopkins Verbal Learning Test, and 7-Minute Screen.

The rationale for the measure underscores important health-related and financial consequences of underrecognized and undertreated mild cognitive impairment that many older adults face, particularly those living in LTC settings. Psychologists have a role (and responsibility) to help close this noted gap in services.

Treatment Plan for the care of the Person with Cognitive Impairment. This Phase II quality measure (currently in the public comment period and therefore considered a draft) piggy backs on the cognitive impairment screening measure. It highlights important high-risk and high-impact areas important to address during the treatment planning and treatment process. That is, the measure emphasizes nutrition, social supports, activity level, cognitive stimulation, and emotional health as key elements to address in the treatment plan for someone with a positive cognitive impairment screen.

As defined by the quality measure, the initial diagnostic interview or examination contains, at a minimum, a history of the presenting problem, pertinent medical, social, and family history, clinical observations, a mental status examination, present evaluation, diagnosis, and recommendations or treatment plan. The areas underscored in the measure for treatment planning purposes were included as they can influence the rate and extent of further cognitive decline.

Nutrition planning includes assessment of the resident's ability to follow the prescribed therapeutic diet (e.g., heart healthy, diabetic) and underscores the risks associated with the use of alcohol and illicit drugs. Social supports are those resources, including material aid, socioemotional support, and information provided by others, that help the older adult cope. Activity-level assessment includes the ability to perform daily physical activity or exercise. Cognitive stimulation includes the ability to engage intellectually.

The measure defines emotional health as a state of emotional and psychological well-being in which an individual is able to use his or her cognitive and emotional capabilities, function in society, and meet the ordinary demands of everyday life. You can see in this measure an overlap of key issues assessed in the MDS described earlier.

In providing evidence in support of this measure, QIP cited the APA's 2004 Guidelines for Psychological Practice with Older Adults, noting that "[o]lder adults can present a broad array of psychological issues requiring clinical attention. Being alert to comorbid physical and mental health problems is a key concept in evaluation and subsequent management of older adults. It is important for psychologists treating older adults to distinguish between mildly impaired cognitive functions and the more serious disorders older adults are at high risk for, to determine the patient's need for assistance with daily activities or treatment" (http://www.usqualitymeasures. org/qm/measures/measurelist.aspx?mid=5).

In support of this measure, Rhew and Shekelle's (2004) paper, Quality Indicators for Preventive Care of the Vulnerable Elder, *RAND Working Paper* (retrieved December 26, 2007, from http://www.rand.org/pubs/working_papers/2005/RAND_ WR178.pdf) was cited. The authors noted that a comprehensive geriatric assessment (CGA) is a multidimensional evaluation, and the medical elements of CGA include special attention to cognition, physical function, affect, visual and hearing impairments, malnutrition, incontinence, and disorders of balance and mobility. Developing this measure for use by psychologists underscores and acknowledges the import role psychologists have in providing a comprehensive biopsychosocial approach to assessment, planning, and treatment services.

Complicated Grief Assessment in Coexisting Disorder. This is a 2009 Phase II (draft) quality measure. The measure looks for documentation of an assessment of a history of losses in the older adult who also presents with depression, anxiety, an adjustment disorder, or substance abuse disorder.

As with the other measures, eligible CPT procedure and ICD-9 CM codes are required to identify patients for inclusion. Eligible procedure codes include 90801, 90802, 90804, 90806, 90808, 90810, 90812, 90814, 90816, 90818, and 90821. There are at least 14 eligible ICD-9 CM diagnosis codes for depression, seven anxiety codes, 11 adjustment disorder codes, and over 30 substance abuse codes appropriate for use with this measure.

By way of definition, the measure notes that "normal grief" symptoms gradually begin fading within 6 months or so of a loss, but the "symptoms of complicated grief get worse or linger for months or even years. Complicated grief is like being in a chronic, heightened state of mourning" (http://www.usqualitymeasures.org/qm/measures/measurelist.aspx?mid=5).

Loss is not limited to death. The measure accounts for this and defines loss more broadly to include not only loss through death but also relationship loss, and/or loss of physical health. The measure acknowledges that "[b]ereaved people are at higher risk of depressive syndromes, sleep disruption, substance abuse and suicide attempts; and the costs, in terms of health care utilization, are substantial." This quality measure suggests that underrecognized and undertreated complicated grief is an issue of concern too important to miss.

Initial Assessment for Posttraumatic Stress Disorder (PTSD) in a Coexisting Disorder. This measure seeks to improve the recognition and treatment of posttraumatic disorder. Under- or missed diagnosis of PTSD contributes to poorer outcomes. Citing the Institute of Medicine (2002), QIP argued that missed comorbid diagnoses, such as PTSD in the presence of depression for example, risks the loss of important information that can affect the approach to treatment and thereby treatment outcomes.

This measure represents an improvement opportunity when treating older adults with co-occurring psychiatric disorders. This lack of documentation or narrow focus of clinical attention on solely the primary diagnosis may result in missed treatment opportunities (and loss of important data), risking less than optimal outcomes (in treatment and possibly even in policy development). As in the complicated grief measure, described earlier, major depression is defined more broadly than the DSM IV coding system and includes instead any of the appropriate ICD-9 CM depression codes. Of note, the major depressive episode must not be due to a medical condition, medication, substance abuse, or psychosis.

Please see the Web site for additional definitions and details by going to <http://www.usqualitymeasures.org/qm/measures/measurelist.aspx?mid=5>. Clearly, there is a role here for mental health professionals with advanced training and education in work with older adults.

Assessment of Suicidal Risk in Coexisting Disorders. This (draft) Phase II mental health measure looks at the percentage of patients aged 65 and above with a coexisting disorder (e.g., depression, anxiety, adjustment, substance abuse) who have a documented suicidal risk assessment. Older adults need special consideration of suicide risk, particularly when comorbid conditions such as bereavement, depression, and terminal illness exist. A psychiatric disorder also presents a risk factor for suicide than any other group (http://www.usqualitymeasures.org/qm/measures/measurelist.aspx?mid=5). Older White males are at highest risk for suicide than are any other group.

In sum, the specific objectives of the PQRI include developing measures that are broadly applicable to physicians and nonphysicians, who provide Part B professional services, and to develop measures for clinical topics that "constitute significant sources of Part B payments and promote a reduction in avoidable hospitalizations, avoidable and chronic disease complications, improved care coordination, improved patient-centered care and services and improved efficiency in the delivery of health care services" (http://www.usqualitymeasures.org/qm/overview/default.aspx).

Summary

Working with older adults, particularly those living in LTC settings, challenges the psychologist to consider the roles and responsibilities of other team members (even if loosely defined as any health care professional providing services to the older adult). Other team members can be an important resource and source of information and visa versa.

Documenting services, according to standards, guidelines, rules, and regulations, helps ensure effective communication among and across disciplines even when keeping in mind important privacy issues. It also helps assure payment for services, as documentation provides the evidence that services provided were medically necessary and appropriate and thus qualify for reimbursement. Documentation creates an important record of services and response to treatment, necessary to measure processes of care and outcomes.

The MDS underlying the nursing home quality improvement program and the Physician's Quality Reporting Initiative reflect the importance of working with older adults, particularly those living in LTC settings, from a biopsychosocial perspective and across disciplines, to identify and treat the varied contributors to an older person's mental health and well-being. There is ample opportunity and great need for well-trained and educated health care professionals to work together to improve the health care quality our nation's seniors receive.

References

American Nurses Association. (1980). *Gerontological nursing: The positive difference in healthcare for older adults*. Kansas City: American Nurses Association.

American Nurses Association. (1982a). *A challenge for change: The role of gerontological nursing*. Kansas City: American Nurses Association.

American Nurses Association. (1982b). *A statement on the scope of gerontological nursing practice*. Kansas City: American Nurses Association.

American Psychological Association (APA). (1995). *Principles for recognition of proficiencies in professional psychology*. Washington, DC: Author.

American Psychological Association (APA). (2004). Guidelines for psychological practice with older adults. *American Psychologist, 59*(4), 236–260.

American Psychological Association (APA). (2007). Record keeping guidelines. *American Psychologist, 62*(9), 993–1004.

Archival Description of Clinical Geropsychology. (n.d.). Retrieved February 2, 2007 from http://www.apa.org/crsppp/gero.html

Arnold, M. (2008). Polypharmacy and older adults: A role for psychology and psychologists. *Professional Psychology Research and Practice, 9(3)*, 283–289.

Cummings, N. A., O'Donohue, W. T., & Ferguson, K. E. (2003). *Behavioral health as primary care: Beyond efficacy to effectiveness.* Cummings Foundation for Behavioral Health: Healthcare Utilization and Cost Series, Volume 6. Reno, NV: Context Press.

APA Division 12, Section II. (1997). *Directory of pre-doctoral internships with clinical geropsychology training opportunities and postdoctoral clinical geropsychology fellowships.* Washington, DC: Author.

HGSA Administrators. (February 2005). Local coverage determination: Psychiatric therapeutic procedures (V-41F). Camp Hill, PA: Author.

Hunsley, J. (2003). Cost-effectiveness and medical cost-offset considerations in psychological service provision. *Canadian Psychology, 44*, 61-73.

Johnson, H. A., Kuder, L. C., & Wellons, K. (1992). Specialization within a generalist social work curriculum. *Journal of Gerontological Social Work, 18*(3-4), 85-98.

Knight, R. and Karel, M. (2006). National Conference on Training in Professional Geropsychology The Pikes Peak Model. *Adult Development and Aging News*, (31)1, pp. 10–12.

Miller, M. S. (ed.). (1997). *Health care choices for today's consumer.* New York: Wiley.

Office of Inspector General. (May 1996). Mental health services in nursing facilities (DHHS Publication No. OEI-02-91-00860). New York Regional Office.

Office of Inspector General. (January 2001). Medicare payments for psychiatric services in nursing homes: A follow-up. (DHHS Publication No. OEI-02-99-00140). New York Regional Office. Office of Inspector General reports are also available at http://www.hhs.gov/oig/oei

Omnibus Budget Reconciliation Act of 1987, PL 100–203 (1993).

Pearson, L. J. (February 2007). The Pearson report. *The American Journal for Nurse Practitioners, 11*(2). http://www.webnp.net/images/ajnp_feb07.pdf

Rhew and Shekelle's paper. (2004). Quality indicators for preventive care of the vulnerable elder. *RAND Working Paper.* (retrieved December 26, 2007 from http://www.rand.org/pubs/working_papers/2005/RAND_WR178.pdf)

Scogin, F. (2007). *Clinical Geropsychology News.* Society of Clinical Geropsychology, APA Division 12, Section II, Volume 14, Number 3.

Strupp, H. H. (1996). The tripartite model and the consumer reports study. *American Psychologist, 51*(10), 1017-1024.

Teri, L., Storandt, M., Gatz, M., Smyer, M., & Stricken, G. (1992). Recommendations from a National Conference on clinical training in psychology: Improving psychological services for older adults. Unpublished manuscript. Washington, DC: American Psychological Association.

http://www.ama-assn.org/ama/pub/category/17432.html
http://www.apa.org/practice/recordkeeping.pdf
http://www.cms.hhs.gov/PQRI
http://www.cms.hhs.gov/DeterminationProcess/04_LCDs.asp#TopOfPage
http://www.empiremedicare.com/newypolicy/policy/L3671_final.htm
http://www.hhs.gov/ocr/index.html
http://www.usqualitymeasures.org/qm/
http://www.usqualitymeasures.org/qm/overview/default.aspx
http://www.usqualitymeasures.org/qm/measures/default.aspx
http://www.usqualitymeasures.org/qm/measures_final/default.aspx
http://www.usqualitymeasures.org/qm/measures/measurelist.aspx?mid=5

Chapter 8
Ethical Issues in Long-Term Care

Michele J. Karel

Introduction

Psychologists practicing in long-term care settings (LTC) must be aware of common ethical concerns that arise in this care environment. The Ethical Principles of Psychologists and Code of Conduct (American Psychological Association, 2002) defines aspirational ethical principles and specific ethical standards to guide all psychological practice, teaching, and research activities.

The general ethical principles are beneficence and nonmalfeasance, fidelity and responsibility, integrity, justice, and respect for people's rights and dignity. These general principles, with which all psychologists should be familiar, are certainly pertinent to psychological practice in LTC. However, psychologists practicing in long-term care will also want to be familiar with the nuances of ethical dilemmas in this setting.

The Standards for Psychological Services in Long-Term Care Facilities (referred to from here on as the Standards; Lichtenberg et al., 1998) highlight and make recommendations regarding five ethical issues: informed consent, confidentiality, privacy, conflict of interest, and advocacy. These ethical issues are reviewed in some detail, followed by a framework that psychologists can use to consider and help resolve a range of ethical dilemmas that can arise in long-term care practice.

Background

Why special ethical considerations in long-term care? Characteristics of both long-term care populations and service settings lead certain ethical issues to come up more frequently than others (Kane & Caplan, 1990; McCullough & Wilson, 1995; Moody, 1992). Individuals requiring long-term care often suffer from neuropsychiatric conditions, such as dementia, that can compromise decision-making capacity (Kim, Karlawish, & Caine, 2002). This fact raises special concerns for informed consent and for promoting the rights and dignity of individuals who may not be able to advocate for themselves.

E. Rosowsky et al. (eds.), *Geropsychology and Long Term Care*,
DOI: 10.1007/978-0-387-72648-9_8, © Springer Science+Business Media, LLC 2009

Further, LTC are typically interdisciplinary in nature, with psychologists collaborating with a treatment team (Ogland-Hand & Zeiss, 2000). Working within the context of a team, often critical to providing care for complex geriatric patients, also has ethical implications regarding patient confidentiality and potential conflicts of interest (e.g., is the resident or team/institution *the client*?). Further, living and working in an institutional setting raises everyday ethical concerns such as constraints on personal choice, conflicts between promoting individual autonomy and protecting safety, and limits on personal privacy.

Ethical issues arise both in direct relation to services offered by the psychologist (e.g., informed consent for psychotherapy, privacy, and confidentiality) and with respect to broader ethical concerns that can arise in long-term care. The psychologist may be asked to consult to help resolve such broader ethical concerns (e.g., assess a resident's capacity to make a particular medical decision, address a diabetic resident's overeating of sweets brought in by the family, etc.). Psychologists should be sensitive to the ethical landscape of the long-term care environment and feel comfortable with strategies to help resolve ethical dilemmas in this setting.

Standards for Psychological Services in Long-Term Care: Ethical Issues

The Standards (Lichtenberg et al., 1998) highlight five ethical issues.

Informed consent: Individual autonomy, or self-determination, is a core value in American medical ethics. Individuals have the right to choose, or refuse, medical or mental health treatments based on their individual values and goals. Valid informed consent for a medical test or treatment must include disclosure, voluntariness, and decision-making competence (Berg, Appelbaum, Lidz, & Parker, 2001; Zehr, 2002).

Disclosure refers to the sharing of sufficient information by a clinician, so that the patient can make a reasonably informed decision (e.g., the diagnosis, prognosis, risks and benefits of treatment alternatives). For informed consent, patients must make decisions freely and voluntarily, without coercion by clinicians, family members, or others, including institutional staff. While clinicians can share treatment recommendations, patients should not feel threatened for choosing a particular course of action (i.e., we will only take care of you if you do "x" but not "y"). Competence refers to the patient's cognitive capacity to make a particular decision. Decision-making capacity is frequently defined as the patient's ability to (1) understand diagnostic and treatment-related information, (2) appreciate how that information relates to oneself personally, (3) reason through the risks and benefits of particular treatment options in light of one's personal values, and (4) express a consistent treatment choice (Appelbaum & Grisso, 1988; Grisso & Appelbaum, 1998). Of note, informed consent issues are pertinent for patients' decisions to participate in clinical research, as well as for clinical decision making (Karlawish & Casarett, 2001).

Informed consent is such a critical issue in the long-term care setting because so many long-term care residents have impaired decisional capacities due to dementia or other chronic illnesses. It is important for psychologists to realize that decision-making capacity is not an *all or nothing* construct. Long-term care residents may have the capacity to make a fairly simple decision such as who is most trusted to be their health care proxy, yet have difficulty in weighing the costs and benefits of a complex medical decision such as whether to proceed with a risky surgical procedure. Psychologists are often called on to help evaluate a resident's capacity to make particular decisions (American Bar Association & American Psychological Association Assessment of Capacity in Older Adults Project Working Group, 2008; Moye, 1999; Moye, Gurrera, Karel, Edelstein, & O'Connell, 2006). It is important that psychologists consider cultural and social contexts in evaluating decision-making capacity (e.g., Karel, 2007).

The Standards provide guidelines for psychologists in obtaining informed consent for psychological services. Distinctions about this process are made for residents who are (1) competent and without significant cognitive impairment, (2) declared legally incompetent and have a legal guardian, or (3) have cognitive impairment and related decisional incapacity but have not been declared legally incompetent. Decisionally capable residents who are referred for psychological testing or treatment services must be informed about "the condition warranting psychological services, what services are to be rendered, and the possible consequences of accepting or refusing services" (p. 125).

Legal guardians must receive the same information regarding legally incompetent patients, and must provide informed consent on behalf of the patient. For patients with impaired capacity, but without a legal guardian, the psychologist provides treatment information to a responsible party (e.g., next of kin), and that caregiver helps the patient make the decision to accept or refuse psychological services. In all cases, the psychologist attempts to explain the rationale for treatment to the patient, to the extent possible given the patient's cognitive status.

The Standards note that consent is not required in emergency, life-threatening situations (consistent with relevant state laws). The Standards also note that psychologists who function as "part of a staff institutional team, privileged by the institution to provide services, and covered by general institutional consent do not need to get separate formal informed consent before implementing treatment" (p. 125). However, it is a good practice in all cases to explain the rationale for treatment and obtain patient assent before proceeding with services (Molinari, McCullough, Coverdale, & Workman, 2006).

An important issue related to informed consent in LTC is the process of advance care planning and the associated use of advance directives. Advance directives are legal mechanisms that aim to protect patient self-determination at a later time, when an individual loses the capacity to make informed medical decisions for himself or herself. Depending on state law, advance directives may be used to designate a durable power of attorney for health care or health care proxy, to make decisions on behalf of a patient if the patient becomes decisionally incapable. They may also document specific treatment preferences, usually related to a patient's wish to accept or refuse life-sustaining interventions in the case of terminal illness, in a living will. Ideally, advance directives are used as part of a broader process of advance care planning, that is, of clarifying and communicating the values and goals that would help to guide future medical decision making (Cantor & Pearlman, 2004; Gillick, Berkman, & Cullen, 1999).

In long-term care practice, psychologists may be asked to assess a patient's capacity to complete an advance directive (e.g., Mezey, Teresi, Ramsey, Mitty, & Bobrowtiz, 2000), to assess whether a patient lacks capacity to the extent that a previously completed advance directive should be enacted (e.g., giving the designated surrogate power to make decisions on behalf of the patient), or, perhaps, to consult on ethical dilemmas arising out of conflicts related to the interpretation of advance directives. Note that there are important differences across ethnic/racial groups in preference regarding end-of-life care, surrogate decision making, and use of advance directives (Kawaga-Singer & Blackhall, 2001; Kwak & Haley, 2005).

Confidentiality: The Standards recognize that, while long-term care residents have the right to confidentiality regarding psychological services, there are multiple valid challenges to strict confidentiality in the long-term care environment (Duffy, 2002; Norris, 2002). In contrast to the outpatient psychotherapy practice setting, psychological services in long-term care are provided in the context of a multidisciplinary team care environment, with shared treatment plans. Often, psychologists can be of most help to long-term care residents by assisting nursing staff to understand, empathize with, and communicate more effectively with those residents. In some cases, withholding psychological/behavioral information about resident functioning from the team could ultimately hurt the resident.

The Standards outline three possible limits to confidentiality: (1) the need to share information "deemed critical to protect the resident from harming self or others," (2) the need to "be consistent with the reporting/charting regulations within which the facility must operate," and (3) the need to "allow for the demands of the psychologist's role as an active member of an institutional treatment team that shares pertinent information with other health professionals" (pp. 125-126). What is critical, according to the Standards, the APA Ethics Code, and HIPAA (Health Insurance Portability and Accountability Act) regulations (APA, 2002), is that patients (or their guardians) are informed about the right to confidentiality as well as the limits to confidentiality as part of the informed consent process. Further, disclosures of confidential information, either with patient consent or in circumstances where disclosure is permitted without consent, should include sharing the minimal amount of information necessary for the purposes of the communication.

Psychologists working in LTC will face frequent dilemmas regarding when and how much information to share with facility staff, residents' family members, and what to document in the medical record. Dr. Margaret Norris writes about "Psychologists' multiple roles in long-term care: Untangling confidentiality quandaries" (2002). Her suggested guidelines for working through these dilemmas are very helpful and will be reviewed here.

1. "The first basic and rather uncomplicated recommendation for psychologists treating long-term care residents is that all patients should be informed at the initiation of services about communication procedures of the facility such as progress notes and treatment team meetings." Psychologists should let patients know that they do communicate with the team, and document services in the medical record, and that together the psychologist and patient will decide what information may be helpful or not helpful to share. The caveat here is that discussions with patients about protection and disclosure of confidential information must be geared toward the cognitive capacity of the patient to understand and appreciate these issues. Further, psychologists may need to introduce this information to initially skeptical individuals who did not themselves seek mental health services, but were referred by the treatment team (as discussed by Duffy, 2002); in these sometimes challenging clinical situations, discussing confidentiality of personal information and the psychologist's potential role as an advocate can help toward building rapport with a reluctant consumer (Norris, 2006, personal communication).

2. "Psychologists should obtain agreements [with their patients] about the type of information that will and will not be communicated to staff." Over the course of psychological assessment or treatment, psychologists will become aware of particular resident fears, preferred coping styles, cognitive strengths vs. deficits, and other psychological conceptualizations of patient behaviors that would be very helpful for the staff to understand. In addition, psychologists frequently hear complaints from long-term care residents about the care they receive (e.g., having to wait too long for assistance, some staff being too rough or rude); sometimes these complaints can be accepted at face value and, at other times, reflect complex perceptual, interpersonal, and systemic issues. Careful consideration must be paid to what will be helpful or not to communicate to staff and, depending upon the resident and the situation, residents can be actively engaged in determining the best course of action.

3. "Psychologists should obtain agreements about what types of information will and will not be communicated with family members." Family members often have concerns about a resident's health, functioning, and care, and their involvement with patient care may range from very constructive and supportive to conflicted and inappropriate (Qualls, 2000; Ogland-Hand & Florsheim, 2002). It is often helpful for psychologists, with permission of the resident, to obtain information from family members, particularly regarding aspects of a resident's history that the resident may not be able to convey accurately herself. In some cases, it may help to include family members directly in psychological treatment. Similarly, it can help to provide family members information about the resident's condition and related capacities to help them optimize their visits and interactions with the resident.

4. "Establish the practice of providing general rather than specific information in medical records." In addition to documenting assessment and treatment services, medical record documentation (i.e., in the general medical record that is shared with other clinical staff) can serve to educate care providers, "to suggest a different

understanding of the patient, to foster empathy for the patient's perspective, and to encourage appropriate interventions by other caregivers" (Moye, 2000, p. 340). Such helpful information can be conveyed without sharing private, intimate details of the patient's story. For example, the psychologist might document that a resident is experiencing PTSD symptoms, and make recommendations for care that will help her to feel safer, without detailing aspects of the personal trauma.

5. "Adopt the 'need to know' principle, i.e., restrict all communication to information that is essential to the care and treatment of the patient." Again, the psychologist will likely come to know many more details about the resident's life and his/her strategies for coping than are necessary to share with staff, family, or the medical record.

Of note, psychologists practicing in long-term care must be aware of HIPAA regulations regarding privacy of protected health information, as many fall under these regulations. These psychologists must provide each patient, or their legal guardian, with a Notice of Privacy Practices at the first contact. Psychologists who work within an organized health care arrangement may be included in a Notice of Privacy Practices given to patients by the long-term care facility, rather than by the psychologist himself/herself. In either case, discussion of confidentiality of health information, and its limits, should occur with each patient. Also, psychologists must comply with medical record documentation standards both of HIPAA and the Medicare Carrier reimbursing psychological services. While HIPAA allows special privacy protection for psychotherapy notes, the following information - often required by insurers - is not protected under psychotherapy notes: counseling session start and stop times; modalities and frequencies of treatments provided; results of clinical tests; and summary of diagnosis, functional status, treatment plan, symptoms, prognosis, and progress to date (APA, 2002). Although these elements, such as summary of functioning and treatment plan, may be released to other caregivers under HIPAA's definition of what is and what is not protected in psychotherapy notes, in fact, these constitute pertinent assessment and treatment data that contribute to the interdisciplinary team's understanding and care of the patient. In other words, although private, confidential information about a patient's psychological treatment may be kept in notes separate from the rest of the medical record, other details about assessment and treatment planning can - and some might argue should – be entered into the medical record.

Privacy: While privacy is typically expected for the delivery of medical or mental health services, privacy is often at a premium in LTC. Nursing home residents have the right to participate in psychological evaluation or psychotherapy services in an environment where others will not be listening in. The Standards advise that psychologists "try to ensure that psychological services are provided in the most private manner possible" (p. 126), while recognizing that assuring privacy can require creativity and, sometimes, less-than-ideal compromise.

Finding a private space to meet, if any exists, can require advance coordination with nursing staff to help the resident get up and out of bed. Sometimes, if weather permits and there is no private indoor space, meetings can be held outdoors. If meeting at a resident's bedside, roommates can be asked to leave. If that is not possible, curtains can be drawn, with the knowledge that others in the room will likely hear; sometimes roommates are so ill that it is clear they are not able to hear or process nearby conversation. Ultimately, it is up to the resident to decide if she is comfortable to meet given the constraints on privacy.

Conflict of interest: Psychologists working in LTC have a primary interest in serving the needs of the LTC resident, but may also have obligations to the interests of the facility, as well as professional self-interest. It is critical in this setting to be clear on "who is the client," while remembering that it is often a combination of the resident, family, and facility. Frequently, however, the patient's values and goals may differ from those of the facility or even the family. A common dilemma for professionals is how to find a balance between promoting patient autonomy (e.g., the patient wishes to continue with unassisted transfers despite repeated falls, or wants to continue living independently despite hospitalizations and failing health) and protecting his or her safety (e.g., providing restraints or move to a setting with more assistance). The interest of the facility, based on the ethical principle of beneficence, or doing what is perceived as best for the patient, is often to protect safety or prevent injury, despite possible constraints on patient autonomy (Kane & Caplan, 1990; Post & Whitehouse, 1995; Powers, 2000). Further, psychologists face conflict of interest when needed services, such as time to meet with the team, are not reimbursable by third-party payers, or when the patient's insurance carrier determines mental health services are no longer medically necessary.

The Standards state: "Psychologists are aware that at times the interests of the facility and the patient may not coincide and make every effort to resolve the conflict in the best interests of the patient" (p. 126). The Standards also warn that psychologists take caution in self-referring for psychological services, and only do so when a clear need is identified and team members are informed.

Further, psychologists make efforts to ensure continuity of psychological care, follow regulations regarding billing third-party payers for reimbursable services, and make care decisions based on patient's best interest rather than reimbursement concerns. Psychologists advocate for policy change when they believe that reimbursement regulations do not serve the best interests of long-term care residents. Establishment of these standards was critical in the face of documented cases of inappropriate billing after Medicare began to reimburse psychologists (Hartman-Stein, 1998).

Advocacy: The Standards highlight advocacy as one special ethical area of concern, with particular attention to advocacy for appropriate use of mental health services in LTC. According to the Standards, psychologists "advocate for the appropriate use of mental health services to reduce excess disability and improve quality of life" and, when such care is not being provided, or provided inappropriately, psychologists work to "educate other care providers to improve the delivery of care in order to be consistent with a biopsychosocial approach to the assessment and treatment of older adults" (p. 126). Mental health services provided in LTC should be consistent with the most current research and clinical guidelines (e.g., Livingston, Johnston, Katona, Paton, & Lyketsos, 2005; Ouslander et al., 2003).

Psychologists may be involved in advocacy activities more broadly, from advocating for an individual patient's rights when appropriate to advocating for health care policy change to improve long-term care services. Many providers in the long-term care setting view part of their role as advocating for patients as needed but, of interest, different providers may be motivated to advocate with different goals. For example, while nurses may typically be trained to advocate for the *best interests* of the patient, social workers and long-term care ombudsmen may view advocating for the patient's autonomy, or self-determination, as the core value (Carter, 2002; Nelson, Allen, & Cox, 2005). As in all ethical dilemmas, several good and compelling principles may be in conflict and lead to different courses of action in any particular situation. Psychologists must be mindful of these ethical principles in choosing when, how, and with what objectives to advocate for patient's rights and/or interests.

Framework for Ethical Decision Making in Long-Term Care

Ethical dilemmas occur when two or more possible courses of action, based on competing values or concerns, come into conflict. Psychologists working in LTC may face ethical dilemmas related directly to psychology practice, as summarized earlier, but may also become aware of, or asked to consult regarding, a range of ethically challenging issues. Presented is a general framework that psychologists can use to consider and facilitate resolution of ethical conflicts that arise in long-term care practice settings. While psychologists are rarely serving as formal ethics consultants, they may find themselves trying to address valid but conflicting concerns of patients, families, team members, and/or facility administrators. This framework is adapted from several sources (Doolittle & Herrick, 1992; Fox, Berkowitz, Chanko, & Powell, 2005; Hanson, Kerkhoff, & Bush, 2005; Holstein & Mitzen, 2001; Karel & Moye, 2006; Moye & Zehr, 2000).

1. Clarify the ethical issue: The first step is to clarify the nature of the ethical conflict or dilemma. Who is concerned and what is each party concerned about? What are the ethical issues involved? Often it can be very helpful simply to define the problem; sometimes clarifying the problem, and the differing perspectives, in a clear, compassionate, and respectful way can help all parties involved realize that they share a number of concerns. In its guidelines for Ethics consultations, the Veterans Health Administration recommends that an ethics question be phrased as follows: "Given [uncertainty or conflict about values], what decisions or actions are ethically justifiable?" or, "Given [uncertainty or conflict about values], is it ethically justifiable to [specific decision or action]?"

For example, this is illustrated in a case where a nursing home resident has demanded to discontinue life-sustaining renal dialysis treatments and the family (including a previously designated health care proxy) and nursing staff are uncomfortable with her decision. "Given the conflict between the [competent] patient's right to choose what medical treatments she wishes to receive or refuse, the family's right to advocate on behalf of their loved one, and the facility's obligation to act in the best interests of the patient, what decisions or actions are ethically justifiable?" This strategy can help to focus a team's efforts at negotiation and decision making. While it can be helpful for the psychologist to be aware of and label the ethical principles or standards at stake in the problem at hand (e.g., autonomy, beneficence, fidelity, informed consent), focusing too broadly on abstract principles in discussions with patients, families, and care teams may not be helpful for resolving ethical dilemmas.

2. Clarify the circumstances, values, and goals of relevant stakeholders. It is important to consider, in a systematic way, the perceived needs and interests, influenced by potentially varying values and goals, of all parties affected by a decision outcome. The patient, various family members (who may be united, or have diverging interests/opinions), staff team members (who, again, may be united, or have diverging interests/opinions), facility administration, other patients who may be affected by decisions (e.g., roommates), and potentially others may have a stake in decision outcomes.

Importantly, the psychologist must also evaluate his/her own values and goals relevant to the decision at hand. An ethical self-assessment is important to help any professional understand his or her potential biases, areas of discomfort, or personal feelings that might interfere with fair resolution of ethical dilemmas (Golden & Sonneborn, 2001). For all parties, what cultural, religious, or personality factors may influence preferences? LTC are often multicultural environments where different values and customs may influence perspectives on a particular issue. For facility and professional staff, what professional, regulatory, or financial concerns may influence preferences? This review may include an evaluation of risk related to different outcomes, and the level of risk that different stakeholders are willing to tolerate.

3. Clarify decision-making authority, related to capacities and relevant policies. In some situations, stakeholders may disagree but ultimately it is clear who has the right to make the decision. For example, a patient who is clearly evaluated to have intact decision-making capacity does have the right to decide whether or not to receive certain medical treatments. Sometimes capable patients can be helped to see why others are concerned and can benefit from additional information, support, and clarification of options. An unpopular decision by a capable patient should certainly be explored further, but as long as there is consensus that the patient has the cognitive and emotional capacity to decide, and has been offered adequate information and support, then the patient has the right to decide. Of course, individual autonomy is not without limit. The patient cannot decide to have services beyond the resources of the institution (e.g., to demand a private room if not available) or the family (e.g., to demand that the daughter quit her job in order to provide care at home), or to ignore the interests of others (e.g., to demand to play television loudly in the middle of the night).

Sometimes, in cases of a marginally capable patient, efforts are needed to help optimize the patient's ability to participate in decision making. Often, the patient's relevant values and goals may be elicited, even in cases of mild to moderate cognitive impairment (Carpenter, Kissel, & Lee, 2007; Karel, Moye, Bank, & Azar, 2007), and this information can help to guide resolution of the dilemma. In the case of an incapable patient, decisional authority may clearly be established through an advance directive; a previously designated health care proxy has the authority to make medical care decisions on behalf of an incompetent patient. However, ethical dilemmas can arise when clinical staff believes that a health care proxy is not acting in the best interests of a patient. Issues of decision-making capacity, surrogate decision making, and institutional policies that define patients' and families' rights and responsibilities are important in this stage of ethical decision making.

4. Intervene to facilitate consideration of ethically justifiable options. A psychologist's communication, interpersonal, negotiation, and analytical skills can all be important in helping to spell out the options, including helping parties to contemplate options not originally considered. What are the pros and cons of different options for meeting the values and goals of the patient, and of other stakeholders? A process of clear and open communication, demonstrating deep respect for the values and concerns of those involved, can often lead to consensus on a choice that is acceptable to all. If it is clear who has decision-making authority in a particular situation, that individual can be helped to make an ethically justifiable decision. It is important to remember that there is rarely one *right* decision; otherwise, there would not be an ethical dilemma.

In the case of the resident refusing her dialysis treatments, many questions need clarification. A psychologist can be of great help in such a situation. Careful evaluation of the resident's decision-making capacity, including the possible influence of emotional and interpersonal issues, as well as understanding the resident's values related to quality of life and her goals of care - and why these considerations may have recently changed - were all critical in this case. A careful evaluation and subsequent intervention found, despite occasional episodes of confusion, a decisionally capable but ambivalent resident who ultimately chose to continue treatment when several interpersonal and comfort concerns were addressed by the family and long-term care facility.

5. Evaluate the resolution of the dilemma. It can be important to determine whether the decision made is addressing concerns of the various stakeholders, depending upon the original dilemma. Many decisions regarding long-term care are not final ones, as circumstances can change over time. For example, a resident may require more or less supervision, or assistance with daily tasks, over time as a disabling condition worsens or improves. Similarly, a resident may have fluctuating decision-making capacity related to acute changes in medical or psychiatric status. Therefore, judgments regarding decision-making authority or levels of care often do need to be reevaluated.

The resident who ultimately chose to continue her dialysis treatment will likely experience changes in decision-making capacity and treatment preferences over time. It is important to confirm who is her preferred health care proxy, should she not be able speak for herself at some future point, and to help her clarify her values and goals regarding at what point would she perceive the burdens of her treatment to outweigh the benefits. With her permission, engaging the family in a discussion of her - and their - current and future concerns can allow the current conflict to serve as an opportunity to respect her values, goals, and dignity in the future.

Summary

Psychologists who practice in LTC should be aware of the ethical landscape of an institutional, multidisciplinary care environment serving many cognitively impaired individuals and, often, their families. Ethical dilemmas range from everyday tensions (e.g., competing goals of promoting autonomy and protecting safety; constraints on personal choice, and privacy, in an institution) to conflicts regarding life-or-death medical decisions. This chapter reviewed the five ethical issues highlighted in the *Standards for Psychological Services in Long-Term Care Facilities* (Lichtenberg et al., 1998): informed consent, confidentiality, privacy, conflict of interest, and advocacy. Addressing these ethical issues requires appreciating the rights of long-term care residents to informed consent, confidentiality, and privacy as well as the very real challenges to these rights.

The psychologist must keep the resident's rights and best interests as the primary considerations, inform residents of limits to confidentiality or conflicts of interest, work with the resident and/or surrogate decision maker and the system to negotiate compromises as needed, and advocate for the resident as appropriate.

A framework for resolving ethical dilemmas in LTC was provided, including five steps: (1) clarify the ethical issue, (2) clarify the circumstances, values, and goals of relevant stakeholders, (3) clarify decision-making authority, related to capacities and relevant policies, (4) intervene to facilitate consideration of ethically justifiable options, and (5) evaluate the resolution of the dilemma. A number of references cited provide further discussion of these issues.

References

American Bar Association & American Psychological Association Assessment of Capacity in Older Adults Project Working Group. (2008). *Assessment of older adults with diminished capacity: A handbook for psychologists.* Washington DC: American Bar Association and American Psychological Association.

American Psychological Association. (2002). Ethical principles of psychologists and code of conduct. *American Psychologist, 57,* 1060–1073.

American Psychological Association Practice Organization. (2002). *Getting ready for HIPAA: A primer for psychologists.* Washington, DC: Author.

Appelbaum, P. S., & Grisso, T. (1988). Assessing patients' capacities to consent to treatment. *New England Journal of Medicine, 319,* 1635–1638.

Berg, J. W., Appelbaum, P. S., Lidz, C. W., & Parker, L. S. (2001). *Informed consent: Legal theory and clinical practice*. New York: Oxford.

Cantor, M. D., & Pearlman, R. A. (2004). Advance care planning in long-term care facilities. *Journal of the American Medical Directors Association, 5*, S72–S80.

Carpenter, B. D., Kissel, E. C., & Lee, M. M. (2007). Preferences and life evaluations of older adults with and without dementia: Reliability, stability, and proxy knowledge. *Psychology and Aging, 22*, 650–655.

Carter, M. W. (2002). Advancing an ethical framework for long-term care. *Journal of Aging Studies, 16*, 57–71.

Doolittle, N. O., & Herrick, C. A. (1992). Ethics in aging: A decision-making paradigm. *Educational Gerontology, 18*, 395–408.

Duffy, M. (2002). Confidentiality and informed consent versus collaboration: Challenges of psychotherapy ethics in nursing homes. In M. P. Norris, V. Molinari, & S. Ogland-Hand (Eds.), *Emerging trends in psychological practice in long-term care* (pp. 277–292). Binghamton, NY: The Haworth Press.

Fox, E., Berkowitz, K. A., Chanko, B. L., & Powell, T. (2005). *Ethics consultation: Responding to ethics concerns in health care*. Washington, DC: National Center for Ethics in Health Care, Veterans Health Administration. Available at http://www.ethics.va.gov/ETHICS/activities/iematerials.asp. Last accessed September 17, 2007.

Gillick, M., Berkman, S., & Cullen, L. (1999). A patient-centered approach to advance medical planning in the nursing home. *Journal of the American Geriatrics Society, 47*, 227–230.

Golden, R. L., & Sonneborn, S. (2001). Ethics in clinical practice with older adults: Recognizing biases and respecting boundaries. In M. B.Holstein & P. B.Mitzen (Eds.), *Ethics in community-based elder care*. New York: Springer.

Grisso, T., & Appelbaum, P. S. (1998). *Assessing competence to consent to treatment*. New York: Oxford.

Hanson, S. L., Kerkhoff, T. R., & Bush, S. S. (2005). *Health care ethics for psychologists: A casebook*. Washington, DC: American Psychological Association.

Hartman-Stein, P. (Ed.). (1998). *Innovative behavioral healthcare for older adults*. San Francisco: Jossey-Bass.

Holstein, M. B., & Mitzen, P. B. (Eds.). (2001). *Ethics in community-based elder care*. New York: Springer.

Kane, R. A., & Caplan, A. L. (1990). *Everyday ethics: Resolving dilemmas in nursing home life*. New York: Singer.

Karel, M. J.(2007). Culture and medical decision making. In S.H. Qualls & M. Smyer (Eds.), *Changes in decision-making capacity in older adults: Assessment and intervention* (pp. 145–174). Hoboken, NJ: Wiley.

Karel, M. J., & Moye, J. (2006). The ethics of dementia caregiving. In S. M. Lobo Prabhu, V. Molinari, & J. W. Lomax (Eds), *Caregiving in dementia: A guide for healthcare professionals* (pp. 261–284). Baltimore: The Johns Hopkins University Press.

Karel, M. J., Moye, J., Bank, A., & Azar, A. (2007). Three methods of assessing values for advance care planning: Comparing persons with and without dementia. *Journal of Aging and Health, 19*, 123–151.

Karlawish, J. H., & Casarett, D. (2001). Addressing the ethical challenges of clinical trials that involve patients with dementia. *Journal of Geriatric Psychiatry and Neurology, 14*, 222–228.

Kawaga-Singer, M., & Blackhall, L. J. (2001). Negotiating cross-cultural issues at the end of life. *Journal of the American Medical Association, 286*, 2993–3001.

Kim, S. Y. H., Karlawish, J. H. T., & Caine, E. D. (2002). Current state of research on decision-making competence of cognitively impaired elderly persons. *American Journal of Geriatric Psychiatry, 10*, 151–165.

Kwak, J., & Haley, W. E. (2005). Current research findings on end-of-life decision making among racially or ethnically diverse groups. *Gerontologist, 45*, 634–641.

Lichtenberg, P. A., Smith, M., Frazer, D., Molinari, V., Rosowsky, E., Crose, R. (1998). Standards for psychological services in long-term care facilities. *The Gerontologist, 38*, 122–127.

Livingston, G., Johnston, K., Katona, C., Paton, J., & Lyketsos, C. G. (2005). Systematic review of psychological approaches to the management of neuropsychiatric symptoms of dementia. *American Journal of Psychiatry, 162*, 1996–2021.

McCullough, L. B., & Wilson, N. L. (1995). *Long-term care decisions: Ethical and conceptual dimensions*. Baltimore: Johns Hopkins University Press.

Mezey, M., Teresi, J., Ramsey, G., Mitty, E., & Bobrowtiz, T. (2000). Decision-making capacity to execute a health care proxy: Development and testing of guidelines. *Journal of the American Geriatrics Society, 48*, 179–187.

Molinari, V., McCullough, L. B., Coverdale, J. H., & Workman, R. (2006). Principles and practice of geriatric assent. *Aging and Mental Health, 10*, 48–54.

Moody, H. R. (1992). *Ethics in an aging society*. Baltimore: Johns Hopkins University Press.

Moye, J. (1999). Assessment of competency and decision making capacity. In P. Lichtenberg (Ed.), *Handbook of assessment in clinical gerontology* (pp. 488–528). New York: Wiley.

Moye, J. (2000). Ethical issues. In V. Molinari (Ed.), *Professional psychology in long term care* (pp. 329–348). New York: Hatherleigh Press.

Moye, J., Gurrera, R. J., Karel, M. J., Edelstein, B., & O'Connell, C. (2006). Empirical advances in the assessment of capacity to consent to medical treatment: Clinical implications and research needs. *Clinical Psychology Review, 26*, 1054–1077.

Moye, J., & Zehr, M. (2000). Resolving ethical challenges for psychological practice in long-term care. *Clinical Psychology: Science and Practice, 7*, 337–344.

Nelson, H. W., Allen, P. D., & Cox, D. (2005). Rights-based advocacy in long-term care: Geriatric nursing and long term-care ombudsmen. *Clinical Gerontologist, 28*, 1–16.

Norris, M. P. (2002). Psychologists' multiple roles in long-term care: Untangling confidentiality quandaries. In M. P. Norris, V. Molinari, & S. Ogland-Hand (Eds.), *Emerging trends in psychological practice in long-term care* (pp. 261–275). Binghamton, NY: The Haworth Press.

Ogland-Hand, S. M., & Florsheim, M. (2002). Family work in a long-term care setting. In M. P. Norris, V. Molinari, & S. Ogland-Hand (Eds.), *Emerging trends in psychological practice in long-term care* (pp. 105–123). Binghamton, NY: The Haworth Press.

Ogland-Hand, S. M., & Zeiss, A. (2000). Interprofessional health care teams. In V. Molinari (Ed.), *Professional psychology in long term care* (pp. 257–277). New York: Hatherleigh Press.

Ouslander, J. G., Bartels, S. J., Beck, C., Beecham, N., Burger, S. G., Clark, T. R. (2003). Consensus statement on improving the quality of mental health care in U.S. nursing homes: Management of depression and behavioral symptoms associated with dementia. *Journal of the American Geriatrics Society, 51*, 1287–1298.

Post, S. G., & Whitehouse, P. J. (1995). Fairhill guidelines on ethics of the care of people with Alzheimer's disease: A clinical summary. *Journal of the American Geriatrics Society, 43*, 1423–1429.

Powers, B. A. (2000). Everyday ethics of dementia care in nursing homes: A definition and taxonomy. *American Journal of Alzheimer's Disease, 15*, 143–151

Qualls, S. (2000). Working with families in nursing homes. In V. Molinari (Ed.), *Professional psychology in long term care* (pp. 329–348). New York: Hatherleigh Press.

Zehr, M. D. (2002). Informed consent in the long term care setting. In M. P. Norris, V. Molinari, & S. Ogland-Hand (Eds.), *Emerging trends in psychological practice in long term care* (pp. 239–260). Binghamton, NY: Haworth Press.

Index

A

Activities of daily living (ADL) assessment, 32–33
Aging, 33
Alcohol abuse assessment, 39
Alzheimer's dementia, 74–75
Anticonvulsant medications
  carbamazepine, 77
  sodium valproate, 78
Antidepressants, 16–17
Antipsychotic medications, 76–77
Anxiolytics, 78

B

Beck Depression Inventory-II rating scale, 36
Behavior problems, 18–19
Benzodiazepines, 78

C

Cholinesterase inhibitors (ChEI), 74–76
Chronic obstructive pulmonary disease (COPD), 18–20
Clinical anxiety inventory, 38
Clinical nurse specialists (CNSs), 89
Clinical social workers (CSWs), 87–88
Clinical services, LTC
  clinical evaluation, 94–95
  discharge planning documentation, 97
  multiple domain outcome measurements, 99–100
  periodic reviews and outcomes measurement, 97–98
  psychological and neuropsychological testing, 95
  psychotherapy considerations, 95–97
  referral, 94
  treatment progress documentation, 98–99

Cognitive disorders, 17–18
Cognitive functioning assessment
  aging, 33
  alcohol abuse and social functioning assessment, 39–40
  environmental assessment, 40
  informant rating scales, 35
  psychological measures, 36–39
  screening instruments administration, 33–34
Coordinated multidisciplinary (interdisciplinary) treatment, 58–59
Current procedural terminology (CPT) codes, 8

D

Dementia. *See also* Behavior problems
  anticonvulsant medications, 77–78
  antipsychotic medications, 76–77
  cholinesterase inhibitors (ChEI), 74–76
  cognitive disorders and, 17–18
Depression
  minor and major problems, 18
  medical assessment and consequences, 73
  selective serotonin reuptake inhibitors (SSRIs), 74

E

Environmental cognitive assessment, 40
Ethical issues, long-term care
  advocacy, 117–118
  confidentiality guidelines, 114–116
  decision-making authority, 119–120
  dilemma resolution evaluation, 120–121
  ethical dilemma, 118–119

informed consent
Ethical issues, long-term care (*cont.*)
   advance directives, 113–114
   decision-making capacity, 112
   long-term care residents distinctions, 113
   privacy and conflict of interest, 116–117
   values and goals, 119

G
Geriatric anxiety inventory (GAI), 38
Geriatric depression scale (GDS), 36–37

H
Hamilton depression rating scale, 36

I
Informant rating scales, 35
Instrumental activities of daily living (IADLs)
   assessment, 32–33
Interdisciplinary approach, 29

L
Local coverage determinations (LCDs), 13–14
Local medical review policies (LMRPs),
   91–92
Long-term care (LTC)
   activities of daily living (ADLs)
      assessment, 32–33
   assessment methods
      behavior observation, 29–30
      clinical interview, 27–28
      interdisciplinary approach, 29
      screening instruments, 28
      self-report and questionnaires, 28–29
   behavior problems, 18–19
   clinical services
      clinical evaluation, 94–95
      ending treatment, 97
      multiple domain outcome
         measurements, 99–100
      periodic reviews and outcomes
         measurement, 97–98
      psychological and neuropsychological
         testing, 95
      psychotherapy notes, 95–97
      referral, 94
      treatment progress documentation,
         98–99
   cognitive functioning assessment
      aging, 33

alcohol abuse and social functioning
     assessment, 39–40
environmental assessment, 40
informant rating scales, 35
psychological measures, 36–39
screening instruments administration,
     33–34
dementia and cognitive disorders, 17–18
depression and anxiety, 18
ethical decision making
   decision-making authority, 119–120
   dilemma evaluation, 120–121
   ethical clarification, 118–119
ethical issues
   advocacy, 117–118
   confidentiality, 114–116
   informed consent, 112–114
   interest conflict, 117
   privacy, 116–117
facility residents typology, 69–71
geropsychology training, 90–91
medical illness, 19–20
medical issues, 68–69
medical record documentation, 91–93
medicare Part B, 13–14
mental health care
   Office of the Inspector General's
     (OIG's), 4–5
   Omnibus Budget Reconciliation Acts
     (OBRAs), 3–4
   reimbursement systems, 5–9
mental health service providers
   clinical nurse specialists (CNSs), 89
   clinical social workers (CSWs), 87–88
   gerontological nursing, 88
   nurse practitioners (NPs), 88–89
   psychiatrists and psychologists, 89
minimum data set collection, 100–101
multiple dimensions, 31–32
multiple informants, 30–31
physicians quality initiation
   clinical depression screening, 104–105
   coexisting disorder, 107–109
   cognitive impairment, 106–107
   physical health status screen, 105–106
professional accountability, 100
psychiatric treatment
   anticonvulsant medications, 77–78
   antipsychotic medications, 76–77
   anxiolytics, 78
   dementia, 74–76
   depression, 72–74
   guiding principles, 71–72
   mental illness, 72

psychologists, 2–3
psychology integration, 79–81
rehabilitation, 20
responsibility assessment
    axis I disorders, 26
    cognitively-impaired individuals,
        25–26
    competent assessments, 24
    content validity, 23
    pharmacological factors, 25
    physical and sensory limitations,
        24–25
    referral questions, 26–27
state laws and regulations, 14–15
treatment plans
    construction, 51–52
    medicare, 50–51
    qualities, 49–50
treatment process
    caseload size and reimbursement
        procedures, 59–60
    coordinated multidisciplinary
        (interdisciplinary) treatment,
        58–59
    demonstrated effectiveness, 56–58
    discontinuing parameters, 60
    ending treatment, 60–61
    optimal treatment methods, 55–56
    psychotherapy, 58

M
Medicare, 50–51
Medicare administrative contractors (MAC),
    6–7, 14
Mental health care
    Office of the Inspector General's (OIG's),
        4–5
    Omnibus Budget Reconciliation Acts
        (OBRAs), 3–4
    reimbursement systems
        billing and coding, 7–8
        consultation requirement, 7
        excluded services, 8–9
        medicare administrative contractors,
            6–7
        outpatient mental health treatment
            limitation, 5–6
Mental health service providers, LTC
    clinical nurse specialists (CNSs), 89
    clinical social workers (CSWs), 87–88
    gerontological nursing, 88
    nurse practitioners (NPs), 88–89
    psychiatrists and psychologists, 89

O
Office of the Inspector General's (OIG's), 4–5
Omnibus Budget Reconciliation Acts
    (OBRAs), 3–4
Optimal treatment methods, 55–56
Outpatient mental health treatment
    limitation, 5

P
Pikes peak model (PPM), 90
Psychiatric treatment
    anxiolytics, 78
    dementia
        anticonvulsant medications, 77–78
        antipsychotic medications, 76–77
        cholinesterase inhibitors (ChEI), 74–76
    depression
        consequences, 72–73
        selective serotonin reuptake inhibitors
            (SSRIs), 73–74
    guiding principles, 71–72
    mental illness, 72
Psychological services
    advocacy, 117–118
    confidentiality, 114–116
    informed consent, 112–114
    interest conflict, 117
    privacy, 116–117
Psychology integration
    assessment, 79–80
    psychological treatment, 80–81
Psychotherapy, 57–58

R
Referral process
    definition, 13
    informal aspects, 15–17
    long-term care (LCD)
        behavior problems, 18–19
        dementia and cognitive disorders,
            17–18
        depression and anxiety, 18
        medical illness, 19–20
        Medicare Part B, 13–14
        rehabilitation, 20
        state laws and regulations, 14–15
    suicide risk, 21
    types, 17
Reimbursement systems
    billing and coding, 7–8
    consultation requirement, 7
    excluded services, 8–9

Reimbursement systems (*cont.*)
    medicare administrative contractors, 6–7
    outpatient mental health treatment
        limitation, 5–6
Responsibility assessment, older adults
    axis I disorders, 26
    cognitively-impaired individuals, 25–26
    competent assessments, 24
    content validity, 23
    pharmacological factors, 25

    physical and sensory limitations, 24–25
    referral questions, 26–27

S
Social functioning assessment, 39–40

T
Therapist authority, 57